The emergence of parallel computers and their potential for the numerical solution of "grand challenge" problems have led to a vast amount of research in domain decomposition methods. These are general flexible methods for the solution of linear or nonlinear systems of equations arising from the discretization of partial differential equations that use underlying properties of the PDEs to obtain fast solutions. The broad family of domain decomposition methods can be said to include many multigrid and multilevel algorithms.

This book presents an easy-to-read discussion of domain decomposition algorithms, their implementation and analysis. The relationship between domain decomposition and multigrid methods is carefully explained at an elementary level, and discussions of the implementation of domain decomposition methods on massively parallel supercomputers are also included. All algorithms are fully described and explained, and a mathematical framework for the analysis and complete understanding of the methods is also carefully developed. In addition, many numerical examples are included to demonstrate the behavior of this important class of numerical methods.

Domain Decomposition is ideal for graduate students about to embark on a career in computational science. It will also be a valuable resource for all those interested in parallel computing and numerical computational methods.

Domain Decomposition

Domain Decomposition

PARALLEL MULTILEVEL METHODS FOR ELLIPTIC PARTIAL DIFFERENTIAL EQUATIONS

Barry F. Smith
Argonne National Laboratory

Petter E. Bjørstad
Universitetet i Bergen

William D. Gropp
Argonne National Laboratory

CAMBRIDGE
UNIVERSITY PRESS

PUBLISHED BY THE PRESS SYNDICATE OF THE UNIVERSITY OF CAMBRIDGE
The Pitt Building, Trumpington Street, Cambridge, United Kingdom

CAMBRIDGE UNIVERSITY PRESS
The Edinburgh Building, Cambridge CB2 2RU, UK
40 West 20th Street, New York NY 10011–4211, USA
477 Williamstown Road, Port Melbourne, VIC 3207, Australia
Ruiz de Alarcón 13, 28014 Madrid, Spain
Dock House, The Waterfront, Cape Town 8001, South Africa

http://www.cambridge.org

First published 1996
First paperback edition 2004

A catalogue record for this book is available from the British Library

Library of Congress Cataloguing-in-Publication Data

Smith, Barry, 1964 –

Domain decomposition : parallel multilevel methods for elliptic
partial differential equations / Barry Smith, Petter Bjørstad,
William Gropp.

p. cm.

Includes bibliographical references (p. –) and index.

ISBN 0 521 49589 X hardback

1. Differential equations, Elliptic – Numerical solutions – Data
processing. 2. Decomposition method – Data processing. 3. Parallel
processing (Electronic computers) I. Bjørstad, Petter E.
II. Gropp, William D. III. Title.
QA377.S586 1996
515′.353 – dc20 95-42362
CIP

ISBN 0 521 49589 X hardback
ISBN 0 521 60286 6 paperback

To Our Parents

Contents

Introduction

THE WIDESPREAD AVAILABILITY of parallel computers and their potential for the numerical solution of difficult-to-solve partial differential equations have led to a large amount of research in domain decomposition methods. Domain decomposition methods are general flexible methods for the solution of linear or nonlinear systems of equations arising from the discretization of partial differential equations (PDEs) that use underlying properties of the PDE to obtain fast solutions. For linear problems, domain decomposition methods can often be viewed as preconditioners for Krylov subspace accelerator techniques such as the conjugate gradient (CG) method or the method of generalized minimum residual (GMRES). For nonlinear problems, they may be viewed as preconditioners for the solution of the linear systems arising from the use of Newton's method or as preconditioners for solvers such as the nonlinear conjugate gradient method. Throughout the text, we consider multigrid and multilevel algorithms as particular examples of the broad class of domain decomposition methods. Chapter 3 is devoted to this important class of methods.

The term **domain decomposition** has slightly different meanings to specialists within the discipline of PDEs:

- In parallel computing, it often means the process of distributing data from a computational model among the processors in a distributed memory computer. In this context, domain decomposition refers to the techniques for decomposing a data structure and can be independent of the numerical solution methods. Data decomposition is perhaps a more appropriate term for this process.
- In asymptotic analysis, it means the separation of the physical domain into regions that can be modeled with different equations, with the interfaces between the domains handled by various conditions (e.g., continuity). In this context, domain decomposition refers to the determination of *which* PDEs to solve.
- In preconditioning methods, domain decomposition refers to the process of subdividing the solution of a large linear system into smaller problems whose solutions can be used to produce a preconditioner (or solver) for the system of equations that results from discretizing the PDE on the entire domain. In this context, domain decomposition refers only to the solution method for the algebraic system of equations arising from the discretization.

Note that all three of these may occur in a single program. For example, asymptotic analysis may be used to identify the PDEs to solve. A preconditioner may be chosen by considering a decomposition that has the properties necessary for both an efficient preconditioner and good parallelism. This book concentrates on domain decomposition methods as preconditioners.

Three of the more important motivations for domain decomposition and multilevel methods are

- ease of parallelization and good parallel performance,
- simplification of problems on complicated geometry, and
- superior convergence properties, nearly $O(1)$ work per degree of freedom.

Although many domain decomposition algorithms have been devised in the past few years, this work has not yet generally led to the widespread adoption of these techniques in the engineering and scientific computing community. One reason is the primitive state of software tools and libraries for parallel computers. Another is the perceived complexity of domain decomposition algorithms. Though a few of the methods are, in fact, rather complicated, most are easily understood and implemented when expressed using matrix notation.

This book is intended to alleviate both of these concerns. It will explain the algorithms in detail and, in addition, discuss how portable, flexible software may be written for parallel computation using domain decomposition techniques. The book is intended for a general audience of engineers, scientists, mathematicians, and computer scientists who are interested in both understanding domain decomposition methods and applying them to their own computational problems. The only mathematics needed to understand most of the material is multivariable calculus and linear algebra plus a basic understanding of finite element discretizations and the iterative solution of linear systems.

Each chapter is divided into three parts. The first part contains a detailed description of the algorithms. This is followed by a discussion of implementation issues. The third section, at the end of each chapter, contains more advanced mathematical ideas, which are never needed to implement the algorithms but are useful to analyze the behavior of the methods and to understand why they work.

Domain decomposition ideas have been applied to a wide variety of problems. We could not hope to include all of these techniques in a single book. Important applications of domain decomposition that we will not discuss include spectral methods, adaptive methods, mixed finite element methods, boundary element methods, fourth order elliptic PDEs, nonlinear problems, computational fluid dynamics, and problems involving operators of mixed type, for instance elliptic and hyperbolic. These are all areas of active research; we refer the readers to the proceedings of the annual domain decomposition meetings, [Glo88a], [Glo88b], [Cha89], [Cha90a], [Cha90b], [Glo91], [Key92a], [Qua94], [Key95], for an extensive survey of recent advances. We hope that the material in this book will provide a solid basis for understanding the more specialized literature. A review article on domain decomposition with less emphasis on the algorithmic details and theory than given here has recently appeared in *Acta Numerica 1994* [Cha94a]. Another source for late breaking news on domain decomposition and multigrid methods is Douglas's MG-Net, which can be accessed via the WWW at `http://na.cs.yale.edu/mgnet/www/mgnet.html`.

Appendix 1 contains a brief introduction to the concepts of preconditioners and Krylov subspace methods. An elementary understanding of these ideas is crucial for appreciating the material in the rest of the book.

The software used for some of the numerical calculations is the Portable Extensionable Toolkit for Scientific computation (PETSc) developed by two of the authors, [Gro94b], and [Smi95]. This is an object-oriented library of software routines and data structures written in C and callable from Fortran 77. The software is introduced in Appendix 2, which also contains information on how to access the software on the Internet. Those readers with no interest in the software aspects may skip Appendix 2.

Readers with a strong interest in parallel computing will note that we do not spend much time on the details of writing parallel programs; the reader is referred to a book on parallel computing for the details (a few starting points are [Fos95], [Gro94a], [Koe94], which discuss the design of parallel programs and the emerging standards, the Message Passing Interface (MPI) and High Performance Fortran (HPF)). We do discuss the issues involved in writing a parallel domain decomposition program, and in the sections on implementation we discuss how to understand, organize, and tune parallel domain decomposition programs.

In this book we have, somewhat artificially, divided domain decomposition algorithms into two classes, those that use *overlapping domains*, which we refer to as *Schwarz methods*, and those that use *nonoverlapping domains*, which we refer to as *substructuring* or *Schur complement* methods. Approximately the first half of the book is devoted to Schwarz methods. These include the classical multigrid algorithms. This is followed by a chapter devoted to substructuring methods. Chapter 5 contains an introduction to the mathematical convergence analysis of domain decomposition algorithms.

An interesting and powerful abstract framework for constructing and analyzing the convergence of domain decomposition and multilevel algorithms has been developed over the last few years. Since this text is intended for a wide audience, we do not begin with the abstract theory, but develop the motivation for it throughout the book and then present the convergence theory in the final chapter. We make no attempt to provide complete analysis of all of the algorithms; instead, we try to provide enough background to prepare the reader for the original sources. Each section has notes and references that direct the reader to more detailed information on topics of particular interest.

In the description and analysis of domain decomposition methods there is a constant interplay between the continuous problem (the PDE and its solution) and discrete approximations to the problem obtained by using, for example, finite element or finite difference methods. In order to prevent confusion we will denote functions by using boldface, for example, $\boldsymbol{u}$, $\boldsymbol{v}$, and discrete functions (vectors in $\mathcal{R}^N$) by using italics, such as u, v. Similarly, operators that act on functions are given in uppercase boldface, for example, $\boldsymbol{A}$, $\boldsymbol{P}$, whereas discrete linear operators (matrices) are represented in italics, such as A, B. Unlike much of the literature on domain decomposition methods we will present all of the algorithms by using matrix notation rather than operator notation. However, for mathematical convenience, the convergence analysis is done with bilinear forms and operator notation.

Acknowledgments

We thank Patrick Ciarlet, Jr., Merete Sofie Eikemo, and Olof Widlund for their many comments on the manuscript. We owe Olof Widlund a special thanks for the use of his extensive bibliography database on domain decomposition methods.

1

One Level Algorithms

PERHAPS the simplest domain decomposition algorithms are the one level, explicitly overlapping Schwarz methods. These methods may often be regarded as generalizations of block Jacobi, block Gauss-Seidel, and line relaxation methods. This chapter is devoted to a thorough discussion of such one level overlapping Schwarz methods. It also introduces much of the notation used in this book, particularly the notation for describing subdomains and the operations on and between them.

The earliest known domain decomposition method was introduced by Schwarz in 1870. Though not originally intended as a numerical method, the classical alternating Schwarz method may be used to solve elliptic boundary value problems on domains that are the union of two subdomains by alternatingly solving the same elliptic boundary problem restricted to the individual subdomains.

1.1 Classical Alternating Schwarz Method

Consider the domain, as shown in Figure 1.1, with $\Omega = \Omega_1 \cup \Omega_2$ on which we wish to solve the linear elliptic PDE

$$\begin{aligned} \boldsymbol{L}\boldsymbol{u} &= f \quad \text{in } \Omega, \\ \boldsymbol{u} &= g \quad \text{on } \partial\Omega. \end{aligned} \tag{1.1}$$

The simplest form of $\boldsymbol{L}$ is minus the Laplacian, $-\Delta$. For simplicity, we restrict our attention to Dirichlet boundary conditions, although more general Neumann and Robin boundary conditions may also be dealt with easily. Let $\partial\Omega$ denote the boundary of Ω and note that the domains Ω, Ω_1, and Ω_2 do not include their boundaries. Also let $\bar{\Omega} = \Omega \cup \partial\Omega$ denote the closure of the domain. The artificial boundaries, Γ_i, are the part of the boundary of Ω_i that is interior to Ω (see Figure 1.1). The rest of the subdomain boundaries are denoted by $\partial\Omega_i \setminus \Gamma_i$: in other words, all of the points on $\partial\Omega_i$ that are not on Γ_i. We will often use this notation to describe the points of the boundary of a subdomain ($\partial\Omega_i$) that are not on an interior boundary (Γ_i).

In order to describe the classical alternating Schwarz method we need to introduce some notations. Let $\boldsymbol{u}_i^n$ denote the approximate solution on $\bar{\Omega}_i$ after n iterations, and $\boldsymbol{u}_1^n|_{\Gamma_2}$ be the restriction of $\boldsymbol{u}_1^n$ to Γ_2; similarly $\boldsymbol{u}_2^n|_{\Gamma_1}$ is the restriction of $\boldsymbol{u}_2^n$ to Γ_1. (More properly, since the $\boldsymbol{u}_i^n$ are not necessarily continuous functions, $\boldsymbol{u}_i^n|_{\Gamma_j}$ is the trace of

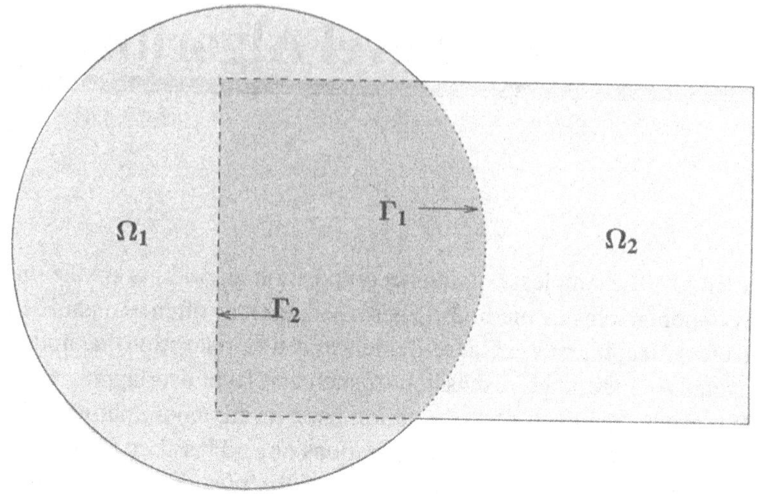

u_i^n on Γ_j, but technical details of this kind are outside the scope of this book and not necessary in order to understand the algorithms.)

The alternating Schwarz method begins by selecting an initial guess, u_2^0, for the values in Ω_2 (actually we only need values along Γ_1). Then, iteratively for $n = 1, 2, 3, \ldots$, one solves the boundary value problem,

$$\begin{aligned} Lu_1^n &= f && \text{in } \Omega_1, \\ u_1^n &= g && \text{on } \partial\Omega_1 \setminus \Gamma_1, \\ u_1^n &= u_2^{n-1}|_{\Gamma_1} && \text{on } \Gamma_1 \end{aligned} \tag{1.2}$$

for u_1^n. This is followed by the solution of the boundary value problem,

$$\begin{aligned} Lu_2^n &= f && \text{in } \Omega_2, \\ u_2^n &= g && \text{on } \partial\Omega_2 \setminus \Gamma_2, \\ u_2^n &= u_1^n|_{\Gamma_2} && \text{on } \Gamma_2. \end{aligned} \tag{1.3}$$

Thus, in each half-step of the alternating Schwarz method we solve the elliptic boundary value problem on the subdomain Ω_i with the given boundary values, g, on the true boundary, $\partial\Omega_i \setminus \Gamma_i$, and the previous approximate solution on the interior boundary Γ_i.

In numerical computation one will, of course, always be working with some finite dimensional discretization of the PDE. We consider two cases, the more general case where the grids (structured or unstructured) do not match between the subdomains and the special case where the grid points (again from structured or unstructured grids) from the two subdomains line up exactly in the overlap region. In order to simplify the notation and decrease the complexity of the text, much of the material in the book will be presented from the point of view of matching grids.

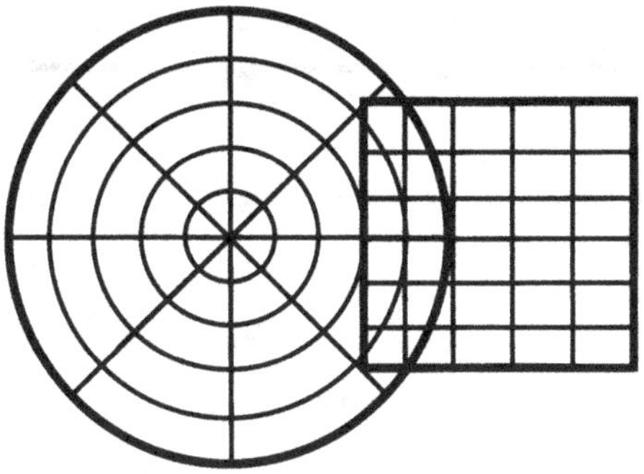

Figure 1.2. Two Nonmatching Grids

1.1.1 General Nonmatching Grids: Alternating Schwarz

Consider the general case where the discretization of Ω is nonconforming: that is, the discretizations of the two domains do not coincide in the overlapping region. A simple example of this might be the discretization of the original example of Schwarz (1.1), using a Cartesian grid for the square and a grid based on polar coordinates in the disk as pictured in Figure 1.2.

In this subsection, we will present the discrete version of the alternating Schwarz method introduced above. In addition, we will show that it is equivalent to a block Gauss-Seidel iteration for a linear system that couples the two subdomains.

Assume that (1.2) and (1.3) have been discretized by using, for instance, finite differences. Then associated with $\boldsymbol{u}_i$ is a vector of coefficients

$$u_i = \begin{pmatrix} u_{\Omega_i} \\ u_{\partial\Omega_i \setminus \Gamma_i} \\ u_{\Gamma_i} \end{pmatrix}.$$

The coefficients $u_{\partial\Omega_i \setminus \Gamma_i}$ are actually known since they are given by the Dirichlet boundary condition. They are kept as "unknowns" in order to simplify the presentation. Associated with the right hand side, $\boldsymbol{f}$, and the boundary values, $\boldsymbol{g}$, are their discrete equivalents, f_i and g_i. The matrix A_i is the discrete form of the operator $\boldsymbol{L}$, restricted to Ω_i. It has three components: the matrix A_{Ω_i} represents the coupling between the interior nodes, $A_{\partial\Omega_i \setminus \Gamma_i}$ represents the coupling between the interior nodes and the true boundary, $\partial\Omega_i \setminus \Gamma_i$, and A_{Γ_i} represents the coupling between the interior nodes and the nodes that lie on the artificial boundary, Γ_i. The matrix A_i is rectangular with one row for each interior node (where it forces the discretization of the PDE to be satisfied) and a column for every node (including the boundary nodes). This can be written as $A_i = (A_{\Omega_i}\ A_{\partial\Omega_i \setminus \Gamma_i}\ A_{\Gamma_i})$. We write this out explicitly for a centered finite difference approximation of the Poisson equation on a rectangular subdomain with two interior points and four points on the artificial boundary; see Figure 1.3.

In order to determine discrete boundary conditions for the internal boundaries Γ_1 and Γ_2, we will need to interpolate values from the other subdomain. Let $\mathcal{I}_{\Omega_j \to \Gamma_i}$ ($i = 1, j = 2$ or $i = 2, j = 1$) denote a discrete operator which interpolates values from the nodes in the interior of Ω_j to the nodes on the curve (or in three dimensions, the

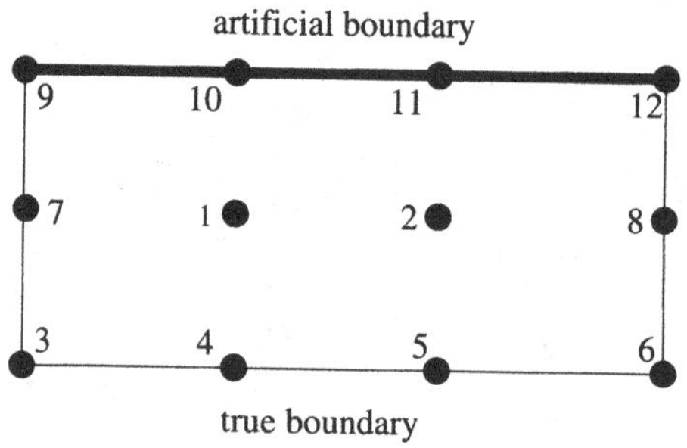

$$\begin{pmatrix} 4 & -1 & \vert & 0 & -1 & 0 & 0 & -1 & 0 & \vert & 0 & -1 & 0 & 0 \\ -1 & 4 & \vert & 0 & 0 & -1 & 0 & 0 & -1 & \vert & 0 & 0 & -1 & 0 \end{pmatrix} \begin{pmatrix} u_1 \\ u_2 \\ \hline u_3 \\ u_4 \\ u_5 \\ u_6 \\ u_7 \\ u_8 \\ \hline u_9 \\ u_{10} \\ u_{11} \\ u_{12} \end{pmatrix} = h^2 \begin{pmatrix} f_1 \\ f_2 \end{pmatrix}$$

Explicit representation of matrix A_i for a simple subdomain and the partial differential equation $-\Delta u = f$ using centered finite differences.

Figure 1.3. Matrix A_i for a Simple Domain

surface) Γ_i. It is possible to use values from $\partial\Omega_j$ in the interpolation process; however in order to simplify the notation we do not consider that case.

The two subproblems in the alternating Schwarz method, (1.2) and (1.3), may be written in a discrete form as

$$\begin{aligned} A_1 u_1^n &= f_1 && \text{in } \Omega_1, \\ u^n_{\partial\Omega_1 \setminus \Gamma_1} &= g_1 && \text{on } \partial\Omega_1 \setminus \Gamma_1, \\ u^n_{\Gamma_1} &= \mathcal{I}_{\Omega_2 \to \Gamma_1} u^{n-1}_{\Omega_2} && \text{on } \Gamma_1, \end{aligned} \tag{1.4}$$

$$\begin{aligned} A_2 u_2^n &= f_2 && \text{in } \Omega_2, \\ u^n_{\partial\Omega_2 \setminus \Gamma_2} &= g_2 && \text{on } \partial\Omega_2 \setminus \Gamma_2, \\ u^n_{\Gamma_2} &= \mathcal{I}_{\Omega_1 \to \Gamma_2} u^{n}_{\Omega_1} && \text{on } \Gamma_2. \end{aligned} \tag{1.5}$$

From these equations we can state the discretized version of the alternating Schwarz method (Algorithm 1.1.1).

Algorithm 1.1.1: Alternating Schwarz Method

- $w_1^0 \leftarrow 0.$
- For $n = 1, \ldots$

  Solve for u_1^n:

$$\begin{aligned} A_1 u_1^n &= f_1 && \text{in } \Omega_1 \\ u^n_{\partial\Omega_1 \setminus \Gamma_1} &= g_1 && \text{on } \partial\Omega_1 \setminus \Gamma_1 \\ u^n_{\Gamma_1} &= w_1^{n-1} && \text{on } \Gamma_1. \end{aligned}$$

  $w_2^n \leftarrow \mathcal{I}_{\Omega_1 \to \Gamma_2} u^n_{\Omega_1}.$

  Solve for u_2^n:

$$\begin{aligned} A_2 u_2^n &= f_2 && \text{in } \Omega_2 \\ u^n_{\partial\Omega_2 \setminus \Gamma_2} &= g_2 && \text{on } \partial\Omega_2 \setminus \Gamma_2 \\ u^n_{\Gamma_2} &= w_2^n && \text{on } \Gamma_2. \end{aligned}$$

  $w_1^n \leftarrow \mathcal{I}_{\Omega_2 \to \Gamma_1} u^n_{\Omega_2}.$

  If $||w_1^n - w_1^{n-1}|| \leq tol_{\Gamma_1}$ and $||w_2^n - w_2^{n-1}|| \leq tol_{\Gamma_2}$ Stop.

  If $||u_1^n - u_1^{n-1}|| \leq tol_{\Omega_1}$ and $||u_2^n - u_2^{n-1}|| \leq tol_{\Omega_2}$ Stop.
- EndFor.

At the termination of the computation, $u^n_{\Omega_1}$ contains the approximate discrete solution for Ω_1, and $u^n_{\Omega_2}$ contains the approximate discrete solution for Ω_2. For the overlap region one is free to use either of the two solutions, or some average, since the two solutions will both converge to the same values as the mesh is refined. Note that the two discrete boundary value problems may be solved by using any direct or iterative procedure.

There are many different approaches to determining when the alternating Schwarz method has converged. Above, we have chosen to require that the approximate solutions in both of the subdomains or on the artificial boundaries change by less than a given tolerance from the previous iteration.

Interestingly, Algorithm 1.1.1 is exactly an application of the block Gauss-Seidel method for a linear system that contains the discretization of both subdomains and the coupling along the artificial boundaries. In order to see this we will rewrite the algorithm in matrix notation.

The two substeps may be rewritten, using the decomposition $A_i = (A_{\Omega_i}\ A_{\partial\Omega_i \setminus \Gamma_i}\ A_{\Gamma_i})$, in the form:

$$
\begin{aligned}
&(A_{\Omega_1}\ A_{\partial\Omega_1\setminus\Gamma_1}\ A_{\Gamma_1})\begin{pmatrix} u^n_{\Omega_1} \\ u^n_{\partial\Omega_1\setminus\Gamma_1} \\ u^n_{\Gamma_1}\end{pmatrix} = f_1 && \text{in } \Omega_1,\\
&u^n_{\partial\Omega_1\setminus\Gamma_1} = g_1 && \text{on } \partial\Omega_1\setminus\Gamma_1,\\
&u^n_{\Gamma_1} = \mathcal{I}_{\Omega_2\to\Gamma_1} u^{n-1}_{\Omega_1} && \text{on } \Gamma_1,\\
&(A_{\Omega_2}\ A_{\partial\Omega_2\setminus\Gamma_2}\ A_{\Gamma_2})\begin{pmatrix} u^n_{\Omega_2} \\ u^n_{\partial\Omega_2\setminus\Gamma_2} \\ u^n_{\Gamma_2}\end{pmatrix} = f_2 && \text{in } \Omega_2,\\
&u^n_{\partial\Omega_2\setminus\Gamma_2} = g_2 && \text{on } \partial\Omega_2\setminus\Gamma_2,\\
&u^n_{\Gamma_2} = \mathcal{I}_{\Omega_1\to\Gamma_2} u^n_{\Omega_1} && \text{on } \Gamma_2.
\end{aligned}
\tag{1.6}
$$

We next move the known quantities to the right hand side and obtain

$$
\begin{aligned}
A_{\Omega_1}u^n_{\Omega_1} &= f_1 - A_{\partial\Omega_1\setminus\Gamma_1}g_1 - A_{\Gamma_1}\mathcal{I}_{\Omega_2\to\Gamma_1}u^{n-1}_{\Omega_2},\\
A_{\Omega_2}u^n_{\Omega_2} &= f_2 - A_{\partial\Omega_2\setminus\Gamma_2}g_2 - A_{\Gamma_2}\mathcal{I}_{\Omega_1\to\Gamma_2}u^{n}_{\Omega_1}.
\end{aligned}
\tag{1.7}
$$

Let $\tilde{f}_i = f_i - A_{\partial\Omega_i\setminus\Gamma_i}g_i$; this is the usual right hand side for Ω_i after one has eliminated the true Dirichlet boundary condition. Then (1.7) may be rewritten as

$$
\begin{aligned}
A_{\Omega_1}u^n_{\Omega_1} &= \tilde{f}_1 - A_{\Gamma_1}\mathcal{I}_{\Omega_2\to\Gamma_1}u^{n-1}_{\Omega_2},\\
A_{\Omega_2}u^n_{\Omega_2} &= \tilde{f}_2 - A_{\Gamma_2}\mathcal{I}_{\Omega_1\to\Gamma_2}u^{n}_{\Omega_1}.
\end{aligned}
$$

This is block Gauss-Seidel for the linear system

$$
\begin{pmatrix} A_{\Omega_1} & A_{\Gamma_1}\mathcal{I}_{\Omega_2\to\Gamma_1} \\ A_{\Gamma_2}\mathcal{I}_{\Omega_1\to\Gamma_2} & A_{\Omega_2}\end{pmatrix}\begin{pmatrix} u_{\Omega_1} \\ u_{\Omega_2}\end{pmatrix} = \begin{pmatrix}\tilde{f}_1 \\ \tilde{f}_2\end{pmatrix}.
\tag{1.8}
$$

Note that even if the two submatrices A_{Ω_1} and A_{Ω_2} are symmetric, the above linear system (1.8) will, in general, not be symmetric, since $A_{\Gamma_1}\mathcal{I}_{\Omega_2\to\Gamma_1} \neq (A_{\Gamma_2}\mathcal{I}_{\Omega_1\to\Gamma_2})^T$.

The discrete alternating Schwarz method can also (by adding and subtracting $u^{n-1}_{\Omega_i}$ in (1.7)) be written in the form:

$$
\begin{aligned}
u^n_{\Omega_1} &\leftarrow u^{n-1}_{\Omega_1} + A^{-1}_{\Omega_1}(\tilde{f}_1 - A_{\Omega_1}u^{n-1}_{\Omega_1} - A_{\Gamma_1}\mathcal{I}_{\Omega_2\to\Gamma_1}u^{n-1}_{\Omega_2}),\\
u^n_{\Omega_2} &\leftarrow u^{n-1}_{\Omega_2} + A^{-1}_{\Omega_2}(\tilde{f}_2 - A_{\Gamma_2}\mathcal{I}_{\Omega_1\to\Gamma_2}u^{n}_{\Omega_1} - A_{\Omega_2}u^{n-1}_{\Omega_2}).
\end{aligned}
\tag{1.9}
$$

Throughout the book, we will use the notation $u^n \leftarrow (\cdot)$ to indicate that the left hand side is assigned the value of the right hand side.

We are now ready to present the alternating Schwarz method (Algorithm 1.1.2) as a preconditioner for a Krylov subspace method. When the alternating Schwarz method is used as a preconditioner, each application of the preconditioner is one iteration of the alternating Schwarz method with a *zero* initial guess. This corresponds to the initialization in Algorithm 1.1.1 with $w^0_1 = 0$.

The linear system (1.8) could also be solved using a block Jacobi preconditioner instead of block Gauss-Seidel; we will refer to this variant as Jacobi-like alternating

Algorithm 1.1.2: Alternating Schwarz Method as a Preconditioner

- $\tilde{f}_1 \leftarrow f_1 - A_{\partial\Omega_1\backslash\Gamma_1} g_1$.
- $\tilde{f}_2 \leftarrow f_2 - A_{\partial\Omega_2\backslash\Gamma_2} g_2$.
- Solve (1.8) with a Krylov subspace method using

  $v \leftarrow$ MatrixMultiply(u):

  $v_{\Omega_1} \leftarrow A_{\Omega_1} u_{\Omega_1} + A_{\Gamma_1} \mathcal{I}_{\Omega_2 \to \Gamma_1} u_{\Omega_2}$

  $v_{\Omega_2} \leftarrow A_{\Gamma_2} \mathcal{I}_{\Omega_1 \to \Gamma_2} u_{\Omega_1} + A_{\Omega_2} u_{\Omega_2}$.

  $v \leftarrow$ Preconditioner(u):

  $v_{\Omega_1} \leftarrow A_{\Omega_1}^{-1} u_{\Omega_1}$

  $v_{\Omega_2} \leftarrow A_{\Omega_2}^{-1} (u_{\Omega_2} - A_{\Gamma_2} \mathcal{I}_{\Omega_1 \to \Gamma_2} v_{\Omega_1})$.

Schwarz method (see Section 1.1.2). The application of $A_{\Omega_1}^{-1}$ and $A_{\Omega_2}^{-1}$ may also be replaced with some suitable approximate solvers; see Section 1.2.

Implementing boundary conditions in a PDE solver can be complicated and error-prone. In the description given above we tacitly assumed that the true Dirichlet boundary conditions were eliminated from the linear system in a preprocessing step (during the formation of $\tilde{f}_i$). In addition, we treated the artificial Dirichlet boundary conditions in a similar way. That is, the known boundary conditions, $u_{\partial\Omega_i\backslash\Gamma_i}$ and u_{Γ_i}, are multiplied by the appropriate parts of A_i, $A_{\partial\Omega_i\backslash\Gamma_i}$, and A_{Γ_i} and subtracted from the right hand side.

In many cases it is more convenient to keep the Dirichlet boundary conditions during the entire calculation, for instance, to simplify data-structures. This is certainly possible with alternating Schwarz methods and the resulting algorithms are essentially identical to those given above. This can be seen by rewriting (1.8) as a larger linear system with unknowns, u_{Ω_i}, $u_{\partial\Omega_i\backslash\Gamma_i}$, and u_{Γ_i}.

$$\begin{pmatrix} A_{\Omega_1} & A_{\partial\Omega_1\backslash\Gamma_1} & A_{\Gamma_1} & & & \\ & I & & & & \\ & & I & -\mathcal{I}_{\Omega_2\to\Gamma_1} & & \\ & & & A_{\Omega_2} & A_{\partial\Omega_2\backslash\Gamma_2} & A_{\Gamma_2} \\ & & & & I & \\ -\mathcal{I}_{\Omega_1\to\Gamma_2} & & & & & I \end{pmatrix} \begin{pmatrix} u_{\Omega_1} \\ u_{\partial\Omega_1\backslash\Gamma_1} \\ u_{\Gamma_1} \\ u_{\Omega_2} \\ u_{\partial\Omega_2\backslash\Gamma_2} \\ u_{\Gamma_2} \end{pmatrix} = \begin{pmatrix} f_1 \\ g_1 \\ 0 \\ f_2 \\ g_2 \\ 0 \end{pmatrix}. \tag{1.10}$$

Note that this is identical to (1.8) except that we have not eliminated the known values. Just as (1.8) may be solved with a block Gauss-Seidel or Jacobi method, so may (1.10). At each half-step one must solve the linear systems

$$\begin{pmatrix} A_{\Omega_i} & A_{\partial\Omega_i\backslash\Gamma_i} & A_{\Gamma_i} \\ & I & \\ & & I \end{pmatrix},$$

which correspond exactly to the discrete boundary value problem with given Dirichlet boundary conditions. In our own codes, we often keep the expanded matrix, stored as a general sparse matrix. Since the number of boundary nodes usually is much smaller than the number of interior nodes, the simplification and increased reuse of the software are often worth the overhead of storing and computing with the "identity" elements.

We now present some sample computational results for the alternating Schwarz method on nonmatching grids.

Computational Results 1.1.1: Alternating Schwarz Method

Purpose: Demonstrate the numerical convergence of the alternating Schwarz method on a nontrivial domain. Also to demonstrate the convergence behavior as the mesh is refined but the geometric overlap remains constant.

PDE: The Poisson equation,

$$\begin{aligned} -\Delta \boldsymbol{u} &= xe^y \quad \text{in } \Omega, \\ \boldsymbol{u} &= -xe^y \quad \text{on } \partial\Omega, \end{aligned}$$

and plane strain linear elasticity,

$$\begin{aligned} -\frac{E}{2(1+\nu)}(\Delta \boldsymbol{u} + \frac{1}{1-2\nu}\nabla\nabla\cdot\boldsymbol{u}) &= -\frac{30E(1-\nu)}{(1+\nu)(1-2\nu)}\begin{pmatrix} x^4 \\ y^4 \end{pmatrix} \quad \text{in } \Omega, \\ \boldsymbol{u} &= \begin{pmatrix} x^6 \\ y^6 \end{pmatrix} \quad \text{on } \partial\Omega, \\ E = 10^7, &\quad \nu = .3. \end{aligned}$$

Domain: Union of a square and a circle:

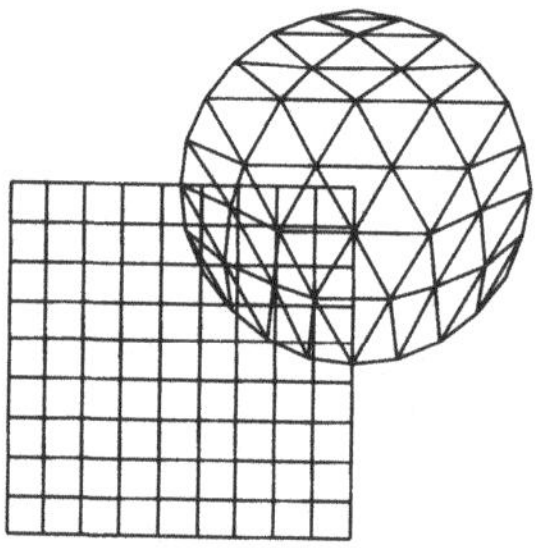

Discretization: For the Laplacian, centered finite difference on the square using a 10×10 grid, and piecewise linear finite elements on the circle, using 64 elements. For elasticity, bilinear finite elements on the square and linear elements on the circle. Piecewise linear interpolation is used from the circle to the boundary of the square while piecewise bilinear is used from the square to the boundary of the circle.

Calculations: Convergence was declared when the first two significant digits of the error no longer changed.

Maximum Error on Each Subdomain for the First Five Iterations

	Error for Laplacian		Error for Elasticity	
Iteration	Ω_1	Ω_2	Ω_1	Ω_2
1	2.7	7.7×10^{-1}	1	4.9×10^{-1}
2	2.1×10^{-1}	7.9×10^{-2}	1.7×10^{-1}	6.3×10^{-2}
3	1.9×10^{-2}	6.6×10^{-3}	1.5×10^{-1}	6.9×10^{-2}
4	1.2×10^{-2}	6.6×10^{-3}	2.1×10^{-1}	8.1×10^{-2}
5	1.2×10^{-2}	6.7×10^{-3}	2.1×10^{-1}	8.7×10^{-2}

Maximum Error After Convergence for Several Levels of Refinement

	No. of Nodes		Laplacian			Elasticity		
Levels	Ω_1	Ω_2	Iter.	Ω_1	Ω_2	Iter.	Ω_1	Ω_2
0	100	45	6	1.2×10^{-2}	6.7×10^{-3}	6	2.2×10^{-1}	8.9×10^{-2}
1	361	153	6	3.1×10^{-3}	2.2×10^{-3}	7	6.3×10^{-2}	2.7×10^{-2}
2	1369	561	7	1.0×10^{-3}	6.9×10^{-4}	9	1.5×10^{-2}	7.3×10^{-3}
3	5329	2145	8	2.7×10^{-4}	2.1×10^{-4}	10	4.0×10^{-3}	2.0×10^{-3}

Discussion: The number of iterations required for convergence to the solution of the algebraic problem is essentially bounded independently of the mesh refinement. In general it depends only on the geometric overlap, which in this case is constant. The convergence of the discrete solution to the continuous solution is somewhere between first and second order. This could be due to the approximation of the circle by straight line segments.

It is natural to ask whether it is beneficial to interpolate the previous approximate solution over the entire overlap region rather than just along the artificial boundary. That is, replace (1.9) by

$$
\begin{aligned}
\tilde{u}_{\Omega_1}^{n-1} &\leftarrow u_{\Omega_1}^{n-1} \quad \text{updated with interpolated values from } u_{\Omega_2}^{n-1},\\
u_{\Omega_1}^{n} &\leftarrow \tilde{u}_{\Omega_1}^{n-1} + A_{\Omega_1}^{-1}(\tilde{f}_1 - A_{\Omega_1}\tilde{u}_{\Omega_1}^{n-1} - A_{\Gamma_1}\mathcal{I}_{\Omega_2\to\Gamma_1}u_{\Omega_2}^{n-1}),\\
\tilde{u}_{\Omega_2}^{n-1} &\leftarrow u_{\Omega_2}^{n-1} \quad \text{updated with interpolated values from } u_{\Omega_1}^{n},\\
u_{\Omega_2}^{n} &\leftarrow \tilde{u}_{\Omega_2}^{n-1} + A_{\Omega_2}^{-1}(\tilde{f}_2 - A_{\Gamma_2}\mathcal{I}_{\Omega_1\to\Gamma_2}u_{\Omega_1}^{n} - A_{\Omega_2}\tilde{u}_{\Omega_2}^{n-1}).
\end{aligned}
\tag{1.11}
$$

However, since the terms $\tilde{u}_{\Omega_i}^{n-1}$ cancel, it is clear that the intermediate values inside Ω_i are irrelevant. See also (1.9).

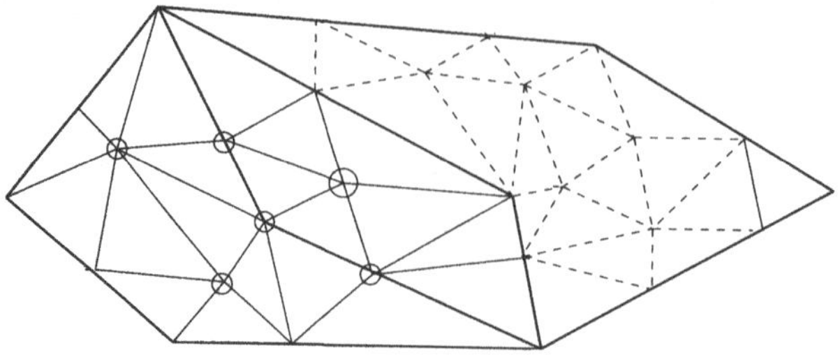

Figure 1.4. Nodes Associated with $u_1^{(1)}$

1.1.2 Matching Grids: Additive and Multiplicative Schwarz

In many applications, it is possible to use grids that match in the overlap region. The Schwarz algorithms may be *changed* and optimized to take advantage of this, to reduce both memory and floating point arithmetic usage. In this section we will modify the algorithms introduced in the previous section to avoid having to deal explicitly with the artificial boundaries. As we will show below, this is achieved by updating the residual in the overlap region before applying the next local solve.

Assume that we have discretized a PDE on the entire grid, obtaining the linear system

$$Au = f, \tag{1.12}$$

where u is a vector whose coefficients represent discrete values of the approximate solution at the mesh points; see Figures 1.4 and 1.5. In the rest of this section, for simplicity, we assume that the nonhomogeneous Dirichlet boundary conditions have already been moved to the right hand side, as in (1.9). Hence f corresponds to the $\tilde{f}$ in the previous section. We order the coefficients in the vector u in the following manner: first the components associated with nodes in the $\Omega_1 \setminus \bar{\Omega}_2$, then those on Γ_2, then those in $\Omega_1 \cap \Omega_2$, then those on Γ_1, and finally those in $\Omega_2 \setminus \bar{\Omega}_1$,

$$u = (u_{\Omega_1\setminus\bar{\Omega}_2}\ u_{\Gamma_2}\ u_{\Omega_1\cap\Omega_2}\ u_{\Gamma_1}\ u_{\Omega_2\setminus\bar{\Omega}_1})^T.$$

Partition the matrix A in two different ways (see Figure 1.6),

$$\begin{pmatrix} A_{\Omega_1} & A_{\Omega\setminus\Omega_1} \\ \cdots & \cdots \end{pmatrix} = \begin{pmatrix} A_{\Omega_1} & A_{\Gamma_1} & A_{\Omega_2\setminus\bar{\Omega}_1} \\ \cdots & \cdots & \cdots \end{pmatrix}$$

and

$$\begin{pmatrix} \cdots & \cdots \\ A_{\Omega\setminus\Omega_2} & A_{\Omega_2} \end{pmatrix} = \begin{pmatrix} \cdots & \cdots & \cdots \\ A_{\Omega_1\setminus\bar{\Omega}_2} & A_{\Gamma_2} & A_{\Omega_2} \end{pmatrix}.$$

In general, A_{Ω_1} and A_{Ω_2} will contain some common entries coming from nodes in the interior of the overlap region. (From Figures 1.4 and 1.5, we see that there is one such common node in Ω_1 and Ω_2 in this example.) Such common nodes will generate equations with identical coefficients in both A_{Ω_1} and A_{Ω_2}. Similarly, u_{Ω_1} and u_{Ω_2} will contain some common values (in the above example, one common value).

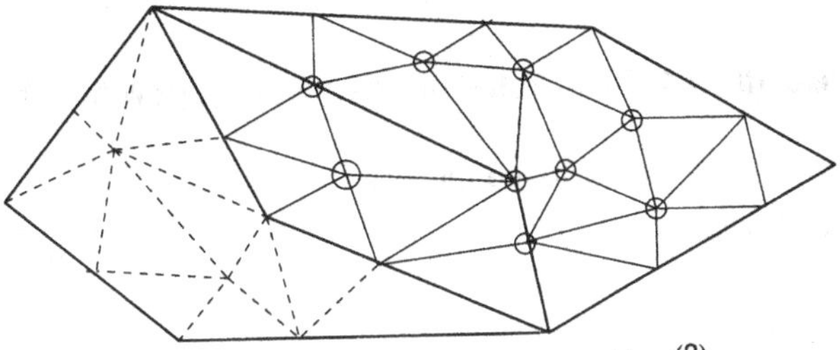

Figure 1.5. Nodes Associated with $u_2^{(2)}$

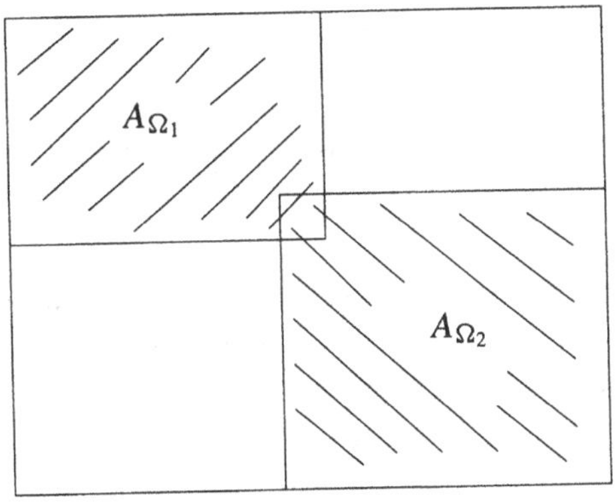

Figure 1.6. Partitioning of Matrix

Recall iteration (1.9)

$$u^n_{\Omega_1} \leftarrow u^{n-1}_{\Omega_1} + A^{-1}_{\Omega_1}(f_1 - A_{\Omega_1}u^{n-1}_{\Omega_1} - A_{\Gamma_1}u^{n-1}_{\Gamma_1}),$$
$$u^n_{\Omega_2} \leftarrow u^{n-1}_{\Omega_2} + A^{-1}_{\Omega_2}(f_2 - A_{\Gamma_2}u^n_{\Gamma_2} - A_{\Omega_2}u^{n-1}_{\Omega_2}).$$

When $A_{\Omega_2\backslash\bar{\Omega}_1} = 0$, and $A_{\Omega_1\backslash\bar{\Omega}_2} = 0$, that is, there is no direct coupling in the matrices between nodes on opposite sides of the artificial boundaries, then $A_{\Gamma_1}u^{n-1}_{\Gamma_1} \equiv A_{\Omega\backslash\Omega_1}u^{n-1}_{\Omega\backslash\Omega_1}$ and similarly $A_{\Gamma_2}u^n_{\Gamma_2} \equiv A_{\Omega\backslash\Omega_2}u^n_{\Omega\backslash\Omega_2}$. Using this observation, we define a new iteration (even when there is direct coupling) by replacing those two terms to obtain

$$u^n_{\Omega_1} \leftarrow u^{n-1}_{\Omega_1} + A^{-1}_{\Omega_1}(f_1 - A_{\Omega_1}u^{n-1}_{\Omega_1} - A_{\Omega\backslash\Omega_1}u^{n-1}_{\Omega\backslash\Omega_1}),$$
$$u^n_{\Omega_2} \leftarrow u^{n-1}_{\Omega_2} + A^{-1}_{\Omega_2}(f_2 - A_{\Omega\backslash\Omega_2}u^n_{\Omega\backslash\Omega_2} - A_{\Omega_2}u^{n-1}_{\Omega_2}).$$

We have now removed the explicit dependence on the artificial boundary from the description of the algorithm. Note that when there is a direct coupling between the two subdomains (for instance, if a high order finite difference method is used), then iteration (1.9) is actually *changed.*

During the second half-step in the interior of Ω_2 we are not using the most recent values of u available to us. In fact, this does not matter. If we rewrite the iteration as

$$u^n_{\Omega_1} \leftarrow u^{n-1}_{\Omega_1} + A^{-1}_{\Omega_1}(f_1 - A_{\Omega_1}u^{n-1}_{\Omega_1} - A_{\Omega\backslash\Omega_1}u^{n-1}_{\Omega\backslash\Omega_1}),$$

Algorithm 1.1.3: Multiplicative Schwarz Method

- Solve (1.12) with a Krylov subspace method using

  $v \leftarrow$ MatrixMultiply(u):

  $v \leftarrow Au$.

  $v \leftarrow$ Preconditioner(u):

  $v_{\Omega_1} \leftarrow A_{\Omega_1}^{-1} u_{\Omega_1}$; $v_{\Omega_2} \leftarrow 0$

  $v_{\Omega_2} \leftarrow A_{\Omega_2}^{-1} R_2(u - Av)$.

$$\tilde{u}_{\Omega_2}^{n-1} \leftarrow u_{\Omega_2}^{n-1} \quad \text{updated with values from } u_{\Omega_1}^{n},$$
$$u_{\Omega_2}^{n} \leftarrow \tilde{u}_{\Omega_2}^{n-1} + A_{\Omega_2}^{-1}(f_2 - A_{\Omega\backslash\Omega_2} u_{\Omega\backslash\Omega_2}^{n} - A_{\Omega_2}\tilde{u}_{\Omega_2}^{n-1}),$$

and if the $A_{\Omega_i}^{-1}$ are solved exactly, then the iterates are the same as those above. Again, only the updated values on the boundary matter; see also (1.11), where we observed a similar effect for the nonmatching grid case.

The terms

$$f_1 - A_{\Omega_1} u_{\Omega_1}^{n-1} - A_{\Omega\backslash\Omega_1} u_{\Omega\backslash\Omega_1}^{n-1}$$

and

$$f_2 - A_{\Omega\backslash\Omega_2} u_{\Omega\backslash\Omega_2}^{n} - A_{\Omega_2} u_{\Omega_2}^{n-1}$$

calculate the new residual in the interior of Ω_i, $i = 1, 2$. Note that this computation need only involve the values of u in Ω_i and near its boundary, since that is where the operator $A_{\Omega_i}^{-1}$ is to be applied.

Algorithm 1.1.3 is the matching version of Algorithm 1.1.2. The duplication of unknowns in the overlapping region has been removed. The matrix R_2 returns the vector of coefficients associated with Ω_2; such matrices will be discussed in more detail below.

In the domain decomposition literature this is known as a **multiplicative Schwarz** method. We will explain the origins of the term multiplicative Schwarz in Section 1.3.1.

We can write the iteration as a two step scheme in a more compact matrix form by also including the part of u that does not change in the first half-step:

$$u^{n+1/2} \leftarrow u^n + \begin{pmatrix} A_{\Omega_1}^{-1} & 0 \\ 0 & 0 \end{pmatrix} (f - Au^n).$$

Similarly, for the second half-step:

$$u^{n+1} \leftarrow u^{n+1/2} + \begin{pmatrix} 0 & 0 \\ 0 & A_{\Omega_2}^{-1} \end{pmatrix} (f - Au^{n+1/2}).$$

Again, in each half-step the residuals need only be updated in the appropriate subdomain.

The subdomain problems in the multiplicative Schwarz methods, unlike in the nonmatching case, must always use **zero** Dirichlet boundary conditions on the artificial boundary. This is because the most recent artificial boundary values are incorporated in the right hand side via $f - Au^n$ or $f - Au^{n+1/2}$. Therefore it would be incorrect to solve the subproblems with nonzero artificial Dirichlet boundary conditions.

We now introduce some additional notation that will be extremely useful throughout the rest of the book.

Let R_i be the rectangular (restriction) matrix that returns the vector of coefficients defined in the interior of Ω_i, that is,

$$u_{\Omega_1} = R_1 u = (I\ 0) \begin{pmatrix} u_{\Omega_1} \\ u_{\Omega \setminus \Omega_1} \end{pmatrix}$$

and

$$u_{\Omega_2} = R_2 u = (0\ I) \begin{pmatrix} u_{\Omega \setminus \Omega_2} \\ u_{\Omega_2} \end{pmatrix}.$$

Here we have reordered the vector u as shown to emphasize that R_i is just a permutation of a rectangular identity matrix. Note that $A_{\Omega_i} = R_i A R_i^T$. The multiplicative Schwarz method may then be written as

$$\begin{aligned} u^{n+1/2} &\leftarrow u^n + R_1^T (R_1 A R_1^T)^{-1} R_1 (f - Au^n), \\ u^{n+1} &\leftarrow u^{n+1/2} + R_2^T (R_2 A R_2^T)^{-1} R_2 (f - Au^{n+1/2}). \end{aligned} \tag{1.13}$$

The R_i matrices are never formed in practice; they are simply introduced to express several different types of algorithms in a similar, concise manner. In implementation it is often convenient to represent R_i as an index set: that is, an integer array containing the global node number of each node in a given subdomain.

Define B_i by $B_i = R_i^T (R_i A R_i^T)^{-1} R_i$. The matrix B_i restricts the residual to one subdomain, solves the problem on the subdomain to generate a correction, and then extends that correction back onto the entire domain. Again, the B_i are never formed explicitly. We can rewrite the two step process in a single step,

$$u^{n+1} \leftarrow u^n + (B_1 + B_2 - B_2 A B_1)(f - Au^n). \tag{1.14}$$

If we interpret this as a Richardson iterative procedure (see Appendix 1.2), we see that the preconditioner B is given by $B_1 + B_2 - B_2 A B_1$. Thus a Krylov subspace accelerator technique can be used to improve the convergence rate of the basic Richardson procedure. An application of the preconditioner B to a vector r is one iteration of the multiplicative Schwarz method with a *zero* initial guess. The following steps will compute $v = Br$:

$$\begin{aligned} v &\leftarrow B_1 r, \\ v &\leftarrow v + B_2 (r - Av). \end{aligned}$$

It is important to realize that the residual, $r - Av$, need not be updated everywhere, only in the region where it affects the action of B_2. In implementations this requires some care. If the discrete operators are stored as sparse matrices we need only apply the rows of A associated with nodes in the interior of Ω_2 to the most recent approximate solution u. Again, this may be done by using an index set for each subdomain.

Even when A is symmetric, the preconditioned conjugate gradient method cannot be used directly since the multiplicative Schwarz preconditioner is not symmetric. However, it is easily symmetrized by including a third substep:

$$\begin{aligned} u^{n+1/3} &\leftarrow u^n + B_1(f - Au^n), \\ u^{n+2/3} &\leftarrow u^{n+1/3} + B_2(f - Au^{n+1/3}), \\ u^{n+1} &\leftarrow u^{n+2/3} + B_1(f - Au^{n+2/3}). \end{aligned} \tag{1.15}$$

This can also be written as a one step method,

$$u^{n+1} \leftarrow u^n + (B_1 + B_2 - B_2AB_1 - B_1AB_2 + B_1AB_2AB_1)(f - Au^n),$$

which does define a symmetric preconditioner that also can be expressed in factored form as

$$B = B_1 + (I - B_1A)B_2(I - AB_1).$$

Again, as in the case of nonmatching grids, multiplicative Schwarz methods may be viewed as generalizations of the block Gauss-Seidel method. The generalization of block Jacobi methods can be written as

$$u^{n+1} \leftarrow u^n + \left(\begin{pmatrix} A_{\Omega_1}^{-1} & 0 \\ 0 & 0 \end{pmatrix} + \begin{pmatrix} 0 & 0 \\ 0 & A_{\Omega_2}^{-1} \end{pmatrix} \right) (f - Au^n).$$

This can also be written as

$$u^{n+1} \leftarrow u^n + (B_1 + B_2)(f - Au^n). \tag{1.16}$$

This is the simplest version of the **overlapping additive Schwarz** method. Again we can interpret the method as a Richardson method, with the preconditioner $B_1 + B_2$. In general this method will not converge; hence in all implementations, the basic Richardson method is accelerated with a Krylov subspace method. See Appendix 1.2.

Another way to derive the additive Schwarz method is to consider the multiplicative Schwarz method written as a one step method; see (1.14). The term $-B_2AB_1$ prevents the parallel application of B_2 and B_1, so we merely drop that term from the preconditioner to obtain (1.16).

Not surprisingly, the convergence rate for the additive Schwarz method is slower than for the multiplicative Schwarz method. In practice, for many problems involving two subdomains, the multiplicative Schwarz method requires approximately one half as many iterations as the additive Schwarz method. This is similar to the classical convergence results for the Jacobi and Gauss-Seidel methods.

Computational Results 1.1.2: Comparison of Schwarz Methods

Purpose: Provide a simple numerical example that readers may duplicate to verify they understand and can implement the algorithms introduced. We also give a simple comparison of convergence rates for these three algorithms.

PDE: The Poisson problem:

$$-\Delta u = xe^y \quad \text{in } \Omega,$$
$$u = -xe^y \quad \text{on } \partial\Omega.$$

Domain: Two by one rectangular region.

Discretization: Centered finite differences with mesh spacing $h = 1/(N+1)$.

Calculations: The same problem was solved for several values of N and several choices for overlap. Convergence was declared when the initial error (in the infinity norm) was reduced by two orders of magnitude. The Krylov subspace method used is GMRES with a restart of 10.

N	Overlap	Alternating	Multiplicative	Additive
11	1	7	2	4
11	2	5	2	3
11	3	4	2	3
21	1	13	3	5
21	2	7	2	4
21	3	5	2	4
31	1	19	3	5
31	2	10	3	5
31	3	7	2	4

Discussion: This table shows several notable features. One of the most important to note is that both the additive and the multiplicative Schwarz methods are relatively insensitive to the amount of overlap, whereas the alternating Schwarz method is quite sensitive. Another notable feature is the factor of nearly 2 difference between the number of iterations for the multiplicative and additive methods. Finally, the relatively poor performance of the alternating Schwarz method relative to the additive and multiplicative methods is also striking, particularly the trend as the problem becomes larger (h smaller).

1.1.3 Corrections as Projections

Define the matrices P_i by

$$P_i = B_i A. \tag{1.17}$$

These operators have important properties that make them the cornerstone of domain decomposition and multigrid/multilevel algorithms.

Note that

$$R_i^T (R_i A R_i^T)^{-1} R_i r = B_i r = B_i A A^{-1} r = P_i e = e_i$$

is a vector in the subspace spanned by the rows of R_i, where e is the error and $r = f - Au^n$ is the residual. In addition, $P_i = P_i^2$ and when A is symmetric, positive definite

$$(P_i x, y)_A = x^T P_i^T A y = x^T A B_i A y = (x, P_i y)_A.$$

Thus, P_i is an **orthogonal projection** (in the A inner product) onto the subspace spanned by the rows of R_i. This means that the correction $e_i = B_i(f - Au^n)$ has a most attractive property: it is the vector *closest* to e in the subspace spanned by the rows of R_i. Here we measure length in the A norm given by $||e||_A = \sqrt{e^T A e}$. We now demonstrate this important property by using elementary calculus and linear algebra.

Any correction that lies in the span of the columns of R_i^T can be written as $e_i = R_i^T w$ for some w. Thus minimizing the "distance" between e_i and e amounts to

$$\min_w ||e - R_i^T w||_A^2$$

or, equivalently,

$$\min_w (e - R_i^T w)^T A (e - R_i^T w).$$

If we take the derivative with respect to the unknown vector w and set it equal to zero we obtain

$$-2 R_i A (e - R_i^T w) = 0$$

or

$$R_i A R_i^T w = R_i A e.$$

Thus the "local" correction is given by

$$e_i = R_i^T w = R_i^T (R_i A R_i^T)^{-1} R_i (f - Au^n).$$

Thus the operator P_i does return the "closest" value to the error.

If $e^n = u^* - u^n$ is the error for the discrete system, then the discrete multiplicative Schwarz method (1.13) may be written as

$$u^{n+1/2} \leftarrow u^n + P_1 e^n,$$
$$u^{n+1} \leftarrow u^{n+1/2} + P_2 e^{n+1/2}.$$

This shows that the algorithm proceeds by projecting the error alternatingly onto the two subspaces $V_i = \text{span}\{R_i^T\}$, and uses these projections as corrections to the approximate solutions.

The procedure of restriction, R_i, (approximately) solving a small linear system, $A_{\Omega_i} u_i = f_i$, and interpolation, R_i^T, lies at the heart of all domain decomposition and multigrid/multilevel algorithms. As indicated above and as we will discuss further in Section 1.6 and in Chapter 5, this procedure calculates the projection of the unknown error onto a subspace and uses the result as a correction term. An effective domain decomposition algorithm will generate the appropriate subspaces that provide a good representation of the solution $\boldsymbol{u}$. This is an important observation that will guide our development of all domain decomposition algorithms.

1.1.4 Other Artificial Boundary Conditions

It is possible to consider more general boundary conditions on the artificial boundary than pure Dirichlet. To make this more concrete we will consider the convection-diffusion equation

$$\begin{aligned} L\boldsymbol{u} = -\Delta \boldsymbol{u} + b(x) \cdot \nabla \boldsymbol{u} &= \boldsymbol{f} \quad \text{in } \Omega, \\ \boldsymbol{u} &= \boldsymbol{g} \quad \text{on } \partial\Omega. \end{aligned}$$

The generalized alternating Schwarz method may be formulated as follows: iteratively for $n = 1, \ldots,$ solve the Robin boundary value problem,

$$\begin{aligned} L\boldsymbol{u}_1^n &= \boldsymbol{f} && \text{in } \Omega_1, \\ \boldsymbol{u}_1^n &= \boldsymbol{g} && \text{on } \partial\Omega_1 \setminus \Gamma_1, \\ \alpha_1 \boldsymbol{u}_1^n + (1-\alpha_1)\frac{\partial \boldsymbol{u}_1^n}{\partial n} &= \alpha_1 \boldsymbol{u}_2^{n-1} + (1-\alpha_1)\frac{\partial \boldsymbol{u}_2^{n-1}}{\partial n} && \text{on } \Gamma_1. \end{aligned}$$

This is followed by the solution of the Robin boundary value problem,

$$\begin{aligned} L\boldsymbol{u}_2^n &= \boldsymbol{f} && \text{in } \Omega_2, \\ \boldsymbol{u}_2^n &= \boldsymbol{g} && \text{on } \partial\Omega_2 \setminus \Gamma_2, \\ \alpha_2 \boldsymbol{u}_2^n + (1-\alpha_2)\frac{\partial \boldsymbol{u}_2^n}{\partial n} &= \alpha_2 \boldsymbol{u}_1^n + (1-\alpha_2)\frac{\partial \boldsymbol{u}_1^n}{\partial n} && \text{on } \Gamma_2. \end{aligned}$$

A matrix formulation of this algorithm may be given by the same techniques outlined in the previous sections. A careful choice of α_i can lead to enhanced convergence. We do not further consider any of the issues related to these types of algorithms.

Notes and References for Section 1.1

- Schwarz [Sch90] originally derived the alternating method to prove existence of a solution to (1.1) in a domain in which there was no known analytic solution.
- P. Lions [Lio78], [Lio88], [Lio89] is responsible for focusing much attention on overlapping Schwarz methods in the 1980s.

- The technique used to symmetrize the preconditioner in (1.15) is the same as that used to convert Gauss-Seidel to a symmetric preconditioner.
- Because of the overlapping blocks it is not possible to write the preconditioners as classical **splittings**, e.g., Varga [Var62]. Tang [Tan92] has defined what he calls generalized Schwarz splittings, which provide a different way of writing the multiplicative Schwarz method so that it looks more like a classical splitting. See also Chapter 3 and Griebel and Oswald [Gri95] for a similar discussion for multilevel methods. Tang also considers the generalized alternating Schwarz method that involves solving local Robin boundary condition problems in each iteration.
- The additive Schwarz preconditioner is automatically suitable for use with the conjugate gradient method as long as B_i and A are symmetric; there is no need to symmetrize the method as there was for the multiplicative Schwarz preconditioner.
- The additive Schwarz method was introduced by Dryja and Widlund in 1987 [Dry87].
- Equation (1.8) can be reduced to a relationship involving only the boundary unknowns by multiplying both sides of the equation by the block diagonal matrix

$$\begin{pmatrix} \mathcal{I}_{\Omega_1\to\Gamma_2} A_{\Omega_1}^{-1} & 0 \\ 0 & \mathcal{I}_{\Omega_2\to\Gamma_1} A_{\Omega_2}^{-1} \end{pmatrix}$$

to obtain the system

$$\begin{pmatrix} I & \mathcal{I}_{\Omega_1\to\Gamma_2} A_{\Omega_1}^{-1} A_{\Gamma_1} \\ \mathcal{I}_{\Omega_2\to\Gamma_1} A_{\Omega_2}^{-1} A_{\Gamma_2} & I \end{pmatrix} \begin{pmatrix} u_{\Gamma_2} \\ u_{\Gamma_1} \end{pmatrix} = \begin{pmatrix} \mathcal{I}_{\Omega_1\to\Gamma_2} A_{\Omega_1}^{-1} \tilde{f}_1 \\ \mathcal{I}_{\Omega_2\to\Gamma_1} A_{\Omega_2}^{-1} \tilde{f}_2 \end{pmatrix}.$$

1.2 Approximate Solvers

In most applications, the Schwarz method is used as a preconditioner for a Krylov subspace method. In these cases, the solution of the subproblems only generates an approximate correction to the global error. It does not seem to make sense then to calculate these approximations too accurately.

Recall that for the matching grid case, $A_{\Omega_i} = R_i A R_i^T$ is the block of A associated with unknowns in the interior of Ω_i. Instead of solving the linear system involving A_{Ω_i} exactly, consider solving it approximately. We will express the approximate solver as $\tilde{A}_{\Omega_i}$ and write the components of the new preconditioner as

$$\tilde{B}_i = R_i^T \tilde{A}_{\Omega_i}^{-1} R_i \qquad T_i = \tilde{B}_i A.$$

Recall that, in the symmetric case, $P_i = B_i A$ was a projection operator onto the subspace spanned by the columns of R_i. Even though T_i is no longer a projection operator, it does, however, approximate the projection. We call it a projectionlike operator.

The standard theory for Krylov subspace based accelerators requires that the preconditioner be the same linear operator in each iteration. This would imply that the operator $\tilde{A}_{\Omega_i}^{-1}$ must be the effect of a linear iteration operator, for instance, a sweep of Gauss-Seidel or the machine precision solution obtained by the preconditioned conjugate gradient method. It could not, however, be the approximate solution obtained by using several steps of a Krylov subspace method, since this is usually not a linear operator. Experience has shown, however, that if the local problems are solved accurately enough, say to several decimal digits, convergence of the outer iteration is not affected much even if the local solver is a nonlinear operator such as a Krylov subspace method. See also the Notes and References for Appendix 1.

It may happen that for large parts of the domain certain terms in the PDE dominate and other terms are much less important. The subdomain solvers can reflect this by simply dropping those terms that are less important and thus obtaining subdomain problems that may be easier to solve, using, for instance, fast solvers. Meanwhile, we do not lose any accuracy in our solution by dropping those terms in the preconditioner because we are still solving the original problem on the entire domain.

1.3 Many Subdomains

Modern parallel computers have from dozens to thousands of independent processors; clearly domain decomposition algorithms using two subdomains would not take advantage of parallel supercomputers. Fortunately, the ideas of alternating and multiplicative Schwarz methods carry over immediately to methods that involve more than two subdomains.

1.3.1 Matching Grids

Consider the domain as pictured in Figure 1.7, where $\Omega_i, i = 1, \ldots, p$ are overlapping subdomains. The **pure multiplicative Schwarz method** involves the following substeps:

$$
\begin{aligned}
u^{n+1/p} &\leftarrow u^n + B_1(f - Au^n),\\
u^{n+2/p} &\leftarrow u^{n+1/p} + B_2(f - Au^{n+1/p}),\\
&\cdots\\
u^{n+1} &\leftarrow u^{n+(p-1)/p} + B_p(f - Au^{n+(p-1)/p}).
\end{aligned}
$$

Recall that $B_i = R_i^T(R_i A R_i^T)^{-1} R_i$ and that $A_i = R_i A R_i^T$ is merely the submatrix of A associated with the domain Ω_i. The residual at each fractional step, $f - Au^{n+(i-1)/p}$, need only be updated in Ω_i using values from Ω_i and its immediate neighbors. As with the two subdomain method, the pure multiplicative Schwarz method may be symmetrized by traversing back through all the subdomains after the final subdomain has been reached.

It is possible to derive an explicit formula for the preconditioner B. Let $e^n = u^* - u^n$ denote the error for the discrete equations, then

$$e^{n+1} = (I - B_p A) \cdots (I - B_1 A)e^n.$$

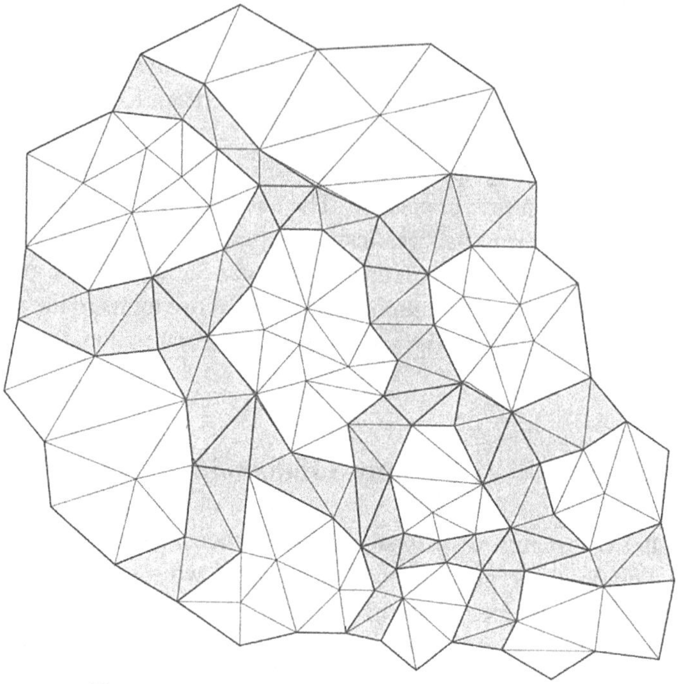

Figure 1.7. Domain with Many Subdomains

Algorithm 1.3.1: Sequential Multiplicative Schwarz Method

- Solve (1.12) with a Krylov subspace method by using
  - $v \leftarrow$ MatrixMultiply(u):
    - $v \leftarrow Au$.
  - $v \leftarrow$ Preconditioner(u):
    - $v \leftarrow B_1 u$
    - For $i = 2, \ldots, p$
      - $v \leftarrow v + B_i(u - Av)$
    - EndFor.

This can be rewritten as

$$\begin{aligned} u^{n+1} &= u^* - (I - B_p A) \cdots (I - B_1 A)(u^* - u^n), \\ &= u^n + (I - (I - B_p A) \cdots (I - B_1 A))(u^* - u^n), \\ &= u^n + (I - (I - B_p A) \cdots (I - B_1 A))A^{-1}(f - Au^n). \end{aligned}$$

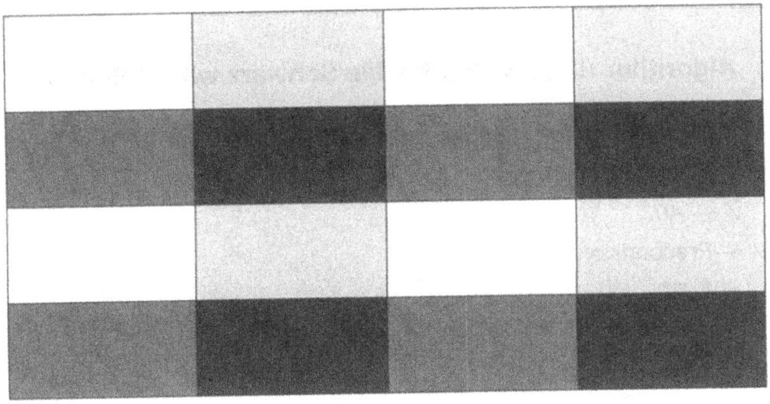

Figure 1.8. Coloring of Subdomains with Four Colors

So we see that the preconditioner, B, is given by

$$\begin{aligned} B &= [I - (I - B_p A) \cdots (I - B_1 A)]A^{-1}, \\ &= A^{-1}[I - (I - AB_p) \cdots (I - AB_1)]. \end{aligned}$$

The above form of the error propagation operator and the explicit formula for the preconditioner in terms of products explain the origins of the term **multiplicative Schwarz**. Note that one would never implement the preconditioner in this form. Sometimes, as above, it is actually confusing to write down explicitly the matrix form of the preconditioner, because of the presence of the A^{-1} term, which, of course, is not needed in the implementation.

The purely multiplicative Schwarz method (see Algorithm 1.3.1) clearly has very little potential for parallelism. This is easily fixed. Note that there are often many subdomains which share no common grid points; see Figure 1.8. Therefore, the solution on all of these subdomains could be updated simultaneously, in parallel.

Define a **coloring** of the subdomains in the following way. For each subdomain, we associate a color, so that no two subdomains which share common points have the same color. Let q be the number of colors used. We can now generate a q-step method,

$$\begin{aligned} u^{n+1/q} &\leftarrow u^n + \sum_{i \in \text{Color}_1} B_i(f - Au^n), \\ u^{n+2/q} &\leftarrow u^{n+1/q} + \sum_{i \in \text{Color}_2} B_i(f - Au^{n+1/q}), \\ &\cdots \\ u^{n+1} &\leftarrow u^{n+(q-1)/q} + \sum_{i \in \text{Color}_q} B_i(f - Au^{n+(q-1)/q}). \end{aligned}$$

In each substep the residual, $f - Au^{n+j/q}$, needs only be updated in the region in which it will affect the action of B_i for that color, that is, in the domains Ω_i for $i \in \text{Color}_{j+1}$. We implement this technique in Algorithm 1.3.2.

In general the convergence rate will depend on the number of colors: the fewer the colors the faster the convergence. However, since coloring the subdomains with the

Algorithm 1.3.2: Multiplicative Schwarz with Coloring

- Solve (1.12) with a Krylov subspace method by using

  $v \leftarrow$ MatrixMultiply(u):

  $v \leftarrow Au$.

  $v \leftarrow$ Preconditioner(u):

  $v \leftarrow \sum_{j \in \text{Color}_1} B_j u$

  For $i = 2, \ldots, q$

  $v \leftarrow v + \sum_{j \in \text{Color}_i} B_j(u - Av)$

  EndFor.

minimum number of colors is difficult, one may use several extra colors to ensure a good load balance of work and data among the processors. This will ideally decrease the convergence rate only slightly.

If the subdomains are logically rectangular, the number of colors needed in two dimensions is usually only four and in three dimensions only eight. In two dimensions, with a square grid of subdomains, one cannot simply use a red-black ordering because all subdomains that share a common vertex must be of a different color, even those with minimal overlap. In practice, the overlap used is never more than a fixed percentage, say 10 to 20 percent, of the width of a subdomain. When there are many subdomains the potential for parallelism is therefore still high. In practice, this means that there should be four or eight times as many subdomains as processors and each processor would have at least four or eight subdomains associated with it, that is, at least one of each color.

To obtain even more parallelism, we can associate one subdomain with each processor and use the additive Schwarz method,

$$u^{n+1} \leftarrow u^n + \sum_i B_i(f - Au^n).$$

There is no guarantee that this simple iteration even converges; in general, it will not. Thus, in practice this method is always accelerated with a Krylov subspace method. The preconditioner for the additive Schwarz method is simply

$$B = \sum B_i.$$

This corresponds to a (generalized) block Jacobi method with (potentially) many overlapping blocks. The term **additive Schwarz** simply refers to the fact that the components of the preconditioner are added together. We implement this in Algorithm 1.3.3.

Algorithm 1.3.3: Additive Schwarz Method

- Solve (1.12) with a Krylov subspace method by using
  - $v \leftarrow$ MatrixMultiply(u):
    - $v \leftarrow Au$.
  - $v \leftarrow$ Preconditioner(u):
    - $v \leftarrow \sum_j B_j u$.

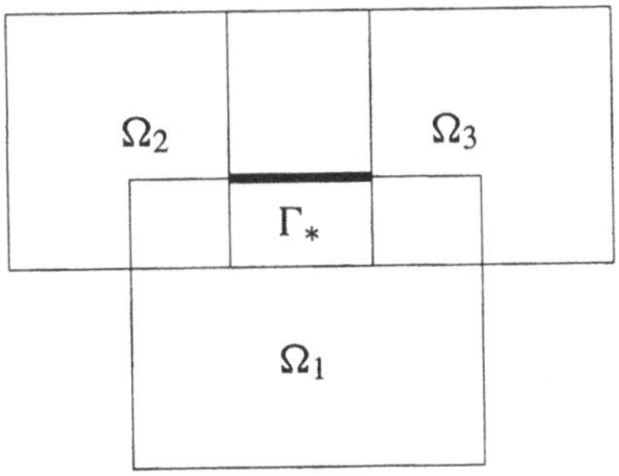

Figure 1.9. Choices of Interpolation Domain

1.3.2 Nonmatching Grids

In order to discuss the case of nonmatching grids, we need to introduce some additional notation. Let $A_i = (A_{\Omega_i}\ A_{\partial\Omega_i\backslash\Gamma_i}\ A_{\Gamma_i})$ and $u_i = (u_{\Omega_i}\ u_{\partial\Omega_i\backslash\Gamma_i}\ u_{\Gamma_i})$ be as defined in Section 1.1.1. In addition, let $u_{\Gamma_{ij}}$ denote the part of u_{Γ_i} in Ω_j where $\Gamma_i = \cup\Gamma_{ij}$. We require that this be a disjoint union, that is, $\Gamma_{ij} \cap \Gamma_{ik} = \emptyset$. We also define $\mathcal{I}_{ij}$ to interpolate from Ω_j to Γ_{ij}. Finally the matrices A_{Γ_i} are partitioned into $A_{\Gamma_{ij}}$.

We then have the following relationships:

$$A_i u_i = f_i,$$

$$u_{\partial\Omega_i\backslash\Gamma_i} = g_i,$$

and

$$u_{\Gamma_{ij}} = \mathcal{I}_{ij} u_{\Omega_j}.$$

Once the Dirichlet boundary values have been eliminated (again, in the actual implementation they need not be), we obtain the linear system,

$$A_{\Omega_i} u_{\Omega_i} + \sum_j A_{\Gamma_{ij}} \mathcal{I}_{ij} u_{\Omega_j} = \tilde{f}_i \qquad \forall i = 1, \ldots, p. \tag{1.18}$$

Algorithm 1.3.4: Alternating Schwarz Method on Many Subdomains

- $\tilde{f}_i \leftarrow f_i - A_{\partial\Omega_i \setminus \Gamma_i} g_i$, for $i = 1, \ldots, p$.
- Solve (1.8) with a Krylov subspace method by using

  $v \leftarrow$ MatrixMultiply(u):

  $v_{\Omega_i} \leftarrow A_{\Omega_i} u_{\Omega_i} + \sum_j A_{\Gamma_{ij}} \mathcal{I}_{ij} u_{\Omega_j}$.

  $v \leftarrow$ Preconditioner(u):

  $v_{\Omega_1} \leftarrow A_{\Omega_1}^{-1} u_{\Omega_1}$

  For $i = 2, \ldots, p$

  $v_{\Omega_i} \leftarrow A_{\Omega_i}^{-1}(u_{\Omega_i} - \sum_{j<i} A_{\Gamma_{ij}} \mathcal{I}_{ij} v_{\Omega_j})$

  EndFor.

The alternating Schwarz method (see Algorithm 1.3.4) then solves this system with a block Gauss-Seidel method, usually as a preconditioner to a Krylov subspace method. Again, as with the matching grid case, the subdomains may be colored and all subdomains of the same color updated simultaneously.

Similarly, this linear system may be solved with block Jacobi preconditioning.

For artificial boundaries that lie in the interior of several subdomains (see Figure 1.9, where u_{Γ_*} may be interpolated from either Ω_2 or Ω_3), there is a choice as to which subdomain to interpolate from. For the alternating Schwarz method, a reasonable choice is to obtain the boundary conditions from the subdomain which has been most recently updated. However, it is also a good idea to interpolate from the center of a domain rather than near its artificial boundary, since the artificial boundary conditions will locally contaminate the approximate solution.

1.4 Convergence Behavior

In this section we discuss the type of convergence one should expect for one level overlapping Schwarz methods. The convergence behavior is essentially what one would expect intuitively.

Consider the domain as in Figure 1.10. Assume that the mesh diameter is $O(h)$ and the subdomains are of diameter $O(H)$ and overlap each other with a width of $O(\delta)$. The number of nodes across a subdomain is $O(H/h)$ and the number of nodes across the entire domain is $O(1/h)$. The number of nodes across an overlap region is $O(\delta/h)$.

For linear elliptic operators where the convection term (lower order derivatives) is not too large, the overlapping Schwarz methods, when used in conjunction with a Krylov subspace method, have the following four properties of convergence behavior:

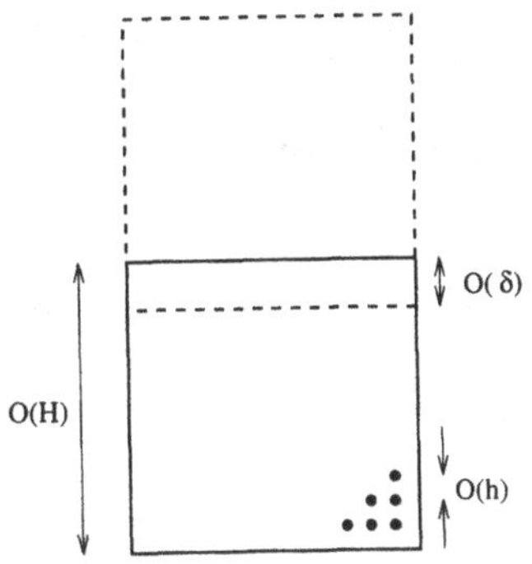

Figure 1.10. Diameter of Subdomains, Overlap, and Elements

- the number of iterations grows as $1/H$,
- if δ is kept proportional to H, the number of iterations is bounded independently of h and H/h,
- the number of iterations for the multiplicative Schwarz method is roughly half of that needed for the additive Schwarz method, and
- convergence is poor for $\delta = 0$ but improves rapidly as the overlap δ is increased.

These can be motivated by considering the basic properties of elliptic PDEs; a more detailed and rigorous analysis is given in Chapter 5.

A key property of elliptic PDEs is that the solution at any point is influenced by all of the boundary values, no matter how distant. This means that for any discrete approximation, the iterative method cannot converge to a correct solution until there has been a chance for data from all the boundary points to affect all of the interior points. This simple observation shows that the point-Jacobi method on an $n \times n$ domain (without any Krylov subspace accelerator) must take at least $O(n)$ iterations, since each iteration has the effect of moving the influence of the boundary points one mesh-width further into the interior.

This same reasoning motivates the first point above: The number of iterations grows like $1/H$ because there are $1/H$ domains along each side, and hence, considering only the interactions in the Schwarz method itself (i.e., ignoring for now the effect of the Krylov subspace acceleration), it will take $1/H$ iterations, for example, for the left boundary to affect the rightmost domain.

To understand the second point above, consider a slightly weaker statement: "If δ is kept proportional to H, and H is kept fixed, the number of iterations is bounded independently of h." All this says is that the behavior of the discrete problem is like that of the continuous problem. That is, if the domains (including their overlap) are fixed, then the number of iterations required is independent of the degree of refinement of the discretization. This result depends on the use of exact solves on the subdomains, though sufficiently good inexact solves may give similar performance.

Note that the convergence is *not* independent of H: changing the number of domains (for single level methods) does affect the convergence. Chapter 2 addresses this issue.

The third point is nothing more than the Schwarz analogue of the result that Gauss-Seidel iteration requires roughly half as many iterations as Jacobi iteration.

The fourth point can be understood in part by recognizing that $\delta = 0$ overlap is essentially block Jacobi (with relatively poor convergence) while large δ approaches a direct solution (each domain covers the entire problem). The only surprise is just how fast the convergence improves as δ increases; typically, δ of just h is far better than δ of zero. This can be understood by looking at the Green's function representation of the solution of an elliptic PDE. The solution $\boldsymbol{u}(x)$ can be written as

$$\boldsymbol{u}(x) = \int G(x, x')\boldsymbol{f}(x')dx'$$

for an appropriate G. The important feature of G is that, while it is non-zero everywhere inside the domain, for many elliptic PDEs it decays rapidly as x' moves away from x. Thus, a small overlap can capture much of the effect of the Green's function.

Computational Results 1.4.1: Studies with Many Subdomains

Purpose: Demonstrate the convergence behavior of one level additive and multiplicative Schwarz methods for many subdomains.

PDE: The Poisson equation,

$$\begin{aligned} -\Delta \boldsymbol{u} &= xe^{y} \quad \text{in } \Omega, \\ \boldsymbol{u} &= -xe^{y} \quad \text{on } \partial\Omega, \end{aligned}$$

with Dirichlet boundary conditions.

Domain: Unit square and an unstructured grid:

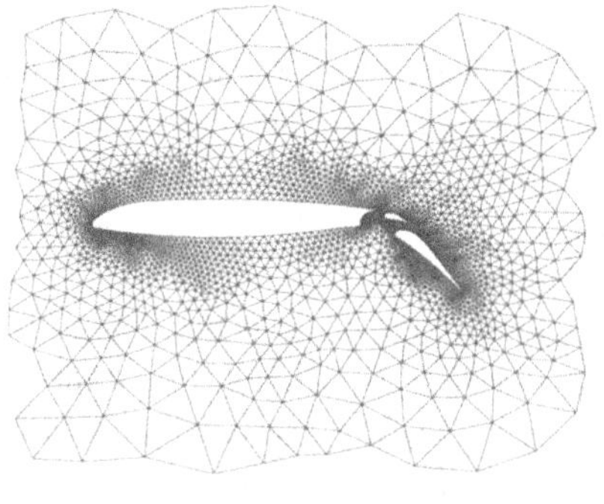

Discretization: Piecewise linear finite elements.

Calculations: For this problem we consider two cases: a fixed number of subdomains with varying overlap and fixed overlap with a varying number of subdomains. We also include a comparison to the classical iterative methods SSOR-GMRES and incomplete LU factorization (ILU); see Appendix 1.1. The Krylov subspace method used is GMRES with a restart of 10.

Discussion: In the case of the problem on the unit square, the decomposition (8, 1) means to divide the domain into eight vertical strips (divide in the x-direction eight times). The decomposition (2, 2) divides the problem into four blocks, cutting both the x and y coordinates in half. The results for this problem show a close relationship between the number of subdomains and the number of iterations (it is particularly close for $N = 33$ and an overlap of 1), as expected from the discussion of convergence.

Iteration Counts for 10^{-2} *Reduction in Error, for Unit Square*

		Multiplicative with Overlap			Additive with Overlap		
N	Decomposition	1	2	4	1	2	4
33	2,1	3	3	2	5	5	4
33	4,1	4	4	3	7	6	5
33	8,1	8	6	4	9	8	6
33	2,2	4	3	3	7	6	4
33	ILU				10		
33	SSOR				18		
65	2,1	4	4	3	7	6	5
65	4,1	6	5	4	10	8	6
65	8,1	8	8	6	16	12	9
65	2,2	5	4	3	10	8	7
65	ILU				19		
65	SSOR				62		

Iteration Counts for 10^{-2} *Reduction in Error, for Unstructured Grid*

Overlap	Domains	Multiplicative	Additive
1	2	4	7
1	4	4	8
1	8	5	10
1	16	5	12
2	2	3	5
2	4	3	7
2	8	4	7
2	16	4	8
4	2	2	3
4	4	2	5
4	8	3	6
4	16	3	6
ILU			13
SSOR			18

This problem was also solved with ILU(0)- and SSOR-preconditioned GMRES. What is most notable here is how quickly the iteration counts grow for these methods when N is doubled.

In the case of the unstructured grid problem, there is no simple decomposition of the domain. In this case, an automatic mesh partitioning code has been used to subdivide the domain. Because of the way in which this partitioning code works, the decomposition into, for example, four subdomains has more in common with a (2, 2) decomposition of the unit square than it does with a (4, 1) decomposition (that is, it is more like two subdomains along each side than four subdomains along one side). This helps explain why the iteration counts are growing slowly with the number of domains. Also notable are the low iteration count, and the nearly two to one ratio of iteration counts between the additive and multiplicative Schwarz methods.

Computational Results 1.4.2: Effects of Ordering on Convection Dominated Problems

Purpose: Demonstrate the effect of the convection term on the convergence behavior of the additive and multiplicative Schwarz methods for many subdomains.

PDE: Convection-diffusion,

$$\begin{aligned} -\Delta \boldsymbol{u} + \beta \cdot \nabla \boldsymbol{u} &= x e^{y} \quad \text{in } \Omega, \\ \boldsymbol{u} &= -x e^{y} \quad \text{on } \partial\Omega, \end{aligned}$$

with Dirichlet boundary conditions. The value β is a vector that gives the direction of convection; in the calculations below, the angle in the descriptions is the angle that this vector makes with the x-axis.

Domain: Unit square.

Discretization: Centered finite differences with upwinding.

Calculations: The figure shows the iteration count as a function of angle for two different decompositions of the domain. This is a polar plot: each point plotted represents a number of iterations (given by the distance from the origin) and an angle for the convection (given by the angle that a line connecting the point with the origin makes with the x-axis). The data were computed at angles of 0, 10, 20, ... degrees, and the lines on the plot connect data at an angle θ with data at an angle of $\theta + 10$.

By comparison, the graphs for the additive Schwarz method are nearly circular and are centered on the origin; in other words, the iteration counts are nearly independent of the angle (or, equivalently, the direction of the vector β).

Discussion: When looking at this graph, it is important to note that it is showing very *good* convergence results for some angles (near 90 degrees) and average results for others, rather than bad results except near 90 degrees. Note that for a range of angles around 90 degrees, it only takes three iterations for the solution to converge. Note also that there is very little dependence in the iteration count on the number of domains for the angle near 90 degrees. This is related to the fact that the first order term is aligned with the decomposition.

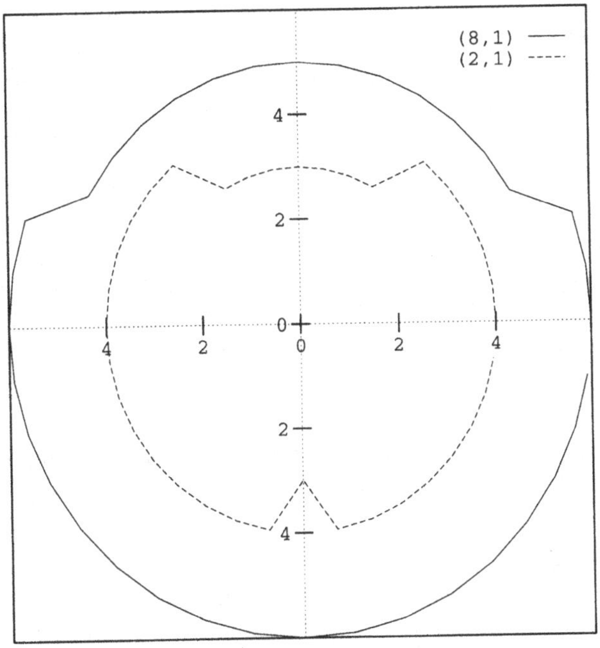

The Plot of Iteration Count as a Function of the Angle for a (2, 1) and an (8,1) Decomposition with an Overlap of 1, a 33 × 33 mesh, and the Multiplicative Schwarz Method. The Magnitude of β is 10. The Krylov Subspace Method used is GMRES with a Restart of 10.

The ability of domain decomposition methods to tune the solution process on each domain is often discussed but rarely used outside the asymptotics community. This simple example shows that, if it is possible to align the domains with the dominant flow direction *locally*, then significant improvements in performance can be achieved.

The results presented here show only two decompositions and only the multiplicative Schwarz method. For larger overlap, just as in the Poisson problem, the iteration count goes down as the amount of overlap is increased. The additive Schwarz method takes more iterations and is somewhat less sensitive to angle than the multiplicative method.

1.5 Implementation Issues

In any discussion of the implementation of domain decomposition methods, it is important to separate the issues that relate to any implementation and those that are specific to a parallel implementation. Many of the issues are independent of the parallelism; a few, such as coloring the domains to achieve better parallelism, are not.

The key to getting efficient use of parallel processors or networks of computers is preserving *data locality*. This is because in most parallel computer architectures the time to move data is far longer than the time it takes to perform arithmetic computations. For instance, the time to move a double precision number between two processors on a certain parallel computer may be more than 50 times as long as it takes to multiply two double precision numbers together. This ratio is often used when analyzing the performance of parallel algorithms. A high ratio requires more local work in order to achieve high efficiency and therefore often implies that each processor needs relatively more local memory. Thus, at least superficially, domain decomposition methods are ideal candidates for efficient implementation on parallel computers.

Less often appreciated is the fact that data locality is a desirable feature for uniprocessors, particularly those with hierarchical cache memories. We can expect better overall performance when each subdomain fits within local memory, whether that is the level 1 or level 2 cache or the processor's main memory in a distributed memory parallel processor.

Before beginning a discussion of particular implementation issues, we make a short digression into some issues of program design. This discussion will help explain our approach and should help the reader understand our implementation choices.

1.5.1 Premature Optimization

In the training of programming for scientific computation the emphasis has historically been on squeezing out every drop of floating point performance for a given algorithm. As the code is written anything that may impede performance, even slightly, is removed. These include subroutine calls, indirect addressing of data, introduction of extra variables, and everything not needed for that particular run. Each piece of code written in this way can perform one type of operation on one type of data and it is usually impossible to reuse the code without a major source code modification.

This practice, however, leads to highly tuned racecarlike software codes: delicate, easily broken and difficult to maintain, but capable of outperforming more user-friendly family cars.

These types of optimizations generally only offer a 5 to 10 percent improvement in the performance of the code but have an enormous cost in the development of general purpose codes. We refer to this as **premature optimization.** Such an approach greatly impedes the development of codes for scientific computation. Unfortunately, it is so strongly ingrained in many of us that it is almost instinctive.

Large scientific computational codes usually spend the vast majority of their time in a few **computational kernels**, such as sparse matrix-vector operations, though a programmer may not readily recognize them as computational kernels. These kernels are what need to be optimized. Pieces of the rest of the code should *only* be optimized if the use of profilers indicates that they are consuming a substantial percentage of the processing time.

1.5.2 Subdomain Solves

The basic building block for implementating one level overlapping Schwarz methods is the subdomain solver. The hope among early domain decomposition workers was that one could write a simple controlling program which would call the old PDE software codes directly to perform the subdomain solves. This turned out to be unrealistic because most PDE software packages are too rigid and inflexible.

The essential feature needed of a subdomain solver is the ability to update its discrete right hand side and boundary conditions and then return the discrete solution at any point in the domain. In addition, if a Krylov subspace method is to be used (which is almost always a good idea), the subdomain solver must also be able to apply the discrete operator to a vector, that is, apply a global matrix vector multiply. Because of the way most PDE solvers handle boundary conditions this is not always easy to accomplish (the handling of boundary conditions is often integrated tightly into the code, through, for example, special handling of the discretization or the matrix elements at the boundary). In addition, the solver must be able to handle many subdomains; many programs are hardwired for a single domain (an example is data in Fortran COMMON blocks).

For matching grids using a multiplicative (Gauss-Seidel–like) version of the overlapping Schwarz method, one must also be able to update the residual only in the region where it is needed, not everywhere.

The easiest way to create this subdomain solver is to use a sparse matrix to describe the discretization (including boundary conditions) and to use a sparse matrix solver to solve the linear system. This approach has the advantage that it remains close to the matrix-based mathematical description used in this book. Once this subdomain solver is working, it is possible to replace the sparse matrix approach with one more closely tuned to the problem (e.g., a regular mesh discretization). Note, however, that this can substantially reduce the flexibility of the code without necessarily increasing the performance of the subdomain solvers. We emphasize this because, whereas most of the computation goes on in the interior of the domains where any regularity in the underlying mesh is easy to describe and exploit, most of the programming involves handling the edges and boundaries.

1.5.3 Data Exchange

Surprisingly, the most complex part of a domain decomposition program to code is often the part that handles gathering the data for the values of a vector restricted to a subdomain. The cause of this is the fact that with overlapping subdomains, there is no simple data structure that can represent both the global vector (all of the unknowns, needed for matrix-vector products and vector-vector products) and the elements of the subdomains. For example, consider a mesh point in the overlap region of four different subdomains. A data-structure representing each subdomain will refer to this mesh point four times; care needs to be taken to ensure that all subdomains get the same value for all mesh points that are in more than one subdomain. In contrast, a data-structure that represents the global vector has no problem representing each element correctly, but it does make it more difficult to extract and set the values of the vector restricted to the subdomain.

One simple approach is to use a direct representation of the R matrices described in the text. These can be implemented as simple sparse matrices; each row of the (rectangular) matrix has a single element of value 1. As before, this operation can be optimized but such optimization should not be taken unless actual performance measurements show that it is necessary.

The approach of using R is clearly simple for the single-processor case; what about the parallel processor case? In fact, the very same approach is appropriate. In the general sparse matrix case, the Krylov subspace method already requires a parallel matrix-vector multiply routine; defining and applying the R matrix is simply another (parallel) matrix-vector multiply.

Again, this operation can be optimized by replacing the general sparse matrix approach with one that knows more about the underlying structure of the problem. This should be viewed as an optimization; the trade-off is programmer time versus (possibly) compute time. Particularly for experimentation with different methods, the advantage in programmer time of using the sparse matrix approach often outweighs any compute-time advantage of using a more tuned approach.

1.5.4 Choosing the Decomposition

An important issue in domain decomposition is choosing the subdomains. For simple tensor-product grids the procedure is straightforward. For unstructured grids it becomes a question of grid partitioning. When applying domain decomposition to parallel computing, there is the goal of decomposing the grid into p pieces each of roughly the same size while minimizing the size of the borders between subdomains, and hence decreasing the communication needed between processors. In addition, ideally the amount of work and storage requirements for each processor should be roughly similar; achieving this is referred to as **load balancing.**

1.5.5 Ordering (Coloring) the Domains

A second issue is the coloring of the resulting subdomains in order to determine which domains should be associated with which processors. The goal is to minimize the number of colors and hence decrease the number of sequential substeps needed in the application of a multiplicative preconditioner.

Both of these issues are discrete combinatorics problems associated with graphs. A **graph** is simply a set of vertices and a set of edges that connect pairs of vertices. We define the **dual graph** of a grid of elements in the following way: Associate a vertex to each element and an edge for each pair of elements that are adjacent. (See Figure 1.11.) The dual graph for a set of subdomains can be defined in a similar manner.

The grid partitioning problem becomes then a question of cutting edges of the graph

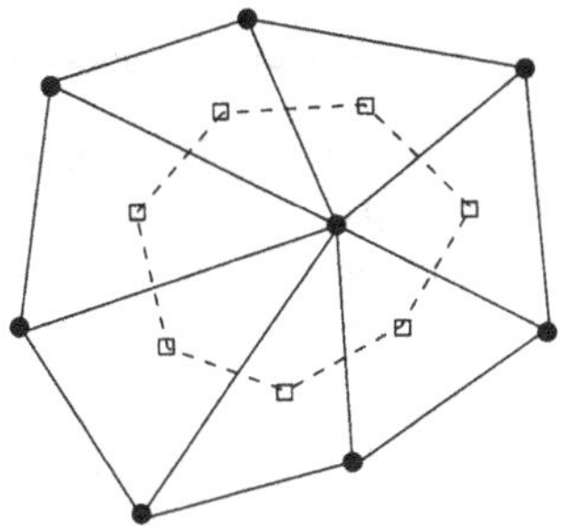

Figure 1.11. A Simple Graph (solid lines) and Its Dual Graph (dashed lines)

until we obtain the desired number of subgraphs (corresponding to subdomains). We would like to minimize the number of cut edges, because each one of these denotes interprocessor communication. In addition, in order to balance the amount of computation for each processor we would like to have approximately the same number of vertices in each subgraph (unknowns in each subdomain).

In the same way, the coloring of subdomains becomes a problem in graph coloring. We would like to color each vertex of the graph so that no other vertex that shares a common edge has the same color.

Both the graph coloring and graph partitioning problems have been studied extensively. They are both NP hard (thus their general solution almost certainly requires an amount of calculation that grows exponentially with the problem size), so rather than try to find exact solutions, researchers have focused on graph coloring and graph partitioning heuristics to find approximate solutions.

1.5.6 Finding the Overlapping Domains

Another complication in implementing an overlapping domain decomposition code is determining the overlapping domains. For a regular mesh, decomposed along the coordinate directions, this is relatively easy. For unstructured meshes, it may seem more difficult. Fortunately, this is yet another graph problem; it can be stated as, given a graph G_1 imbedded within another graph G, compute the graph G_2 that contains G_1 and all of the adjacent vertices. This is a relatively straightforward operation and can be applied recursively to generate any degree of overlap.

1.5.7 Where Do the Graphs Come From?

The graphs discussed above can be computed directly from the sparse-matrix representation of the problem. Each unknown (degree of freedom) is a vertex; there is an edge between two vertices if the corresponding matrix entry is nonzero. That is, if the element a_{ij} of the matrix is nonzero, then there is an edge between vertex v_i and v_j. This allows a one level domain decomposition method to operate entirely with the sparse-matrix representation of the problem, without any additional information about the discretization domains.

Notes and References for Section 1.5

- Graph coloring is a key step in many numerical and nonnumerical problems. Brelaz describes graph coloring in [Bre79]. Some general parallel algorithms are discussed in [Zer90], [Lub86]. Jones and Plassmann discuss a parallel graph coloring method in the context of numerical work in [Jon93].
- There is an extensive literature devoted to the partitioning of meshes for parallel computing. Kernighan and Lin's [Ker70] paper provides a good reference for the general problem without any reference to parallel computing. A sampling of parallel techniques may be found in [Sav91], [Gold92], [Gil87], and [Jon94]. An interesting approach to graph partitioning are the spectral methods; [Pot90] and [Pot92] discuss recent work in this area.
- Hendrickson and Leland have written a comprehensive software package named Chaco [Hen93a] containing advanced graph partitioning algorithms [Hen95], [Hen93b].
- P. Ciarlet, Jr., Lamour, and Smith have considered the effect of the partioning algorithm used on the numerical convergence of the overlapping Schwarz method in [Cia95].

1.6 Variational Formulation

This section develops the ideas introduced in Subsection 1.1.3 further and more abstractly.

In order to analyze the convergence behavior of domain decomposition algorithms we need to introduce a mathematical framework. This is most easily done in the context of Sobolev spaces and Galerkin finite elements. Indeed, it turns out that the basic correction steps calculated in virtually all domain decomposition and multigrid/multilevel algorithms may be viewed as (approximate) orthogonal projections, in some suitable inner product, onto a subspace. This observation makes possible the complete analysis of many domain decomposition and multigrid methods.

In this section we introduce some of the spaces that are needed for the analysis and demonstrate that the corrections obtained in the one level overlapping Schwarz methods are, in fact, projections. These results will be used in Chapter 5 to prove convergence for several standard domain decomposition methods.

Derivations of this kind can usually only be done in the context of Galerkin finite element methods and have not been extended fully for finite difference methods, which in the general case, seem to be more difficult. The algorithms are, however, fully applicable for finite difference methods as well as other discretization techniques.

The manipulations in this section are similar to those previously introduced in the chapter. The difference is that here the operators act on the actual functions $\boldsymbol{u}$ rather than on the vector of coefficients, u. For simplicity, we consider only homogeneous Dirichlet boundary conditions, that is, $\boldsymbol{g} = 0$. After an integration by parts (actually an application of a Green's formula), (1.1) can be expressed in weak form as

$$a(\boldsymbol{u}, \boldsymbol{v}) = f(\boldsymbol{v}) \qquad \forall \boldsymbol{v} \in H_0^1(\Omega).$$

The space $L^2(\Omega)$ consists of all functions $\boldsymbol{u}$ such that $\int_\Omega \boldsymbol{u}^2 < \infty$. The space $H_0^1(\Omega)$

consists of all functions in $L^2(\Omega)$ whose first derivatives are also in $L^2(\Omega)$, the trace of which (for continuous functions this merely means the value of the function) is zero on the boundary of Ω.

In order to understand this more concretely, consider the special case where $\boldsymbol{L} = -\Delta$, that is, (1.1) is given by

$$-\Delta \boldsymbol{u} = f \quad \text{in } \Omega,$$
$$\boldsymbol{u} = 0 \quad \text{on } \partial\Omega.$$

Then, always assuming $\boldsymbol{f} \in L^2(\Omega)$,

$$a(\boldsymbol{u}, \boldsymbol{v}) = \int_\Omega \nabla u \cdot \nabla v$$

and

$$f(\boldsymbol{v}) = \int_\Omega f\boldsymbol{v}.$$

The form $a(\boldsymbol{u}, \boldsymbol{v})$ is called **bilinear** since it is linear in both $\boldsymbol{u}$ and $\boldsymbol{v}$. For general, second order elliptic operators $\boldsymbol{L}$ the bilinear form need not be symmetric. If the operator is self-adjoint and the resulting bilinear form is uniformly elliptic (i.e., $a(\boldsymbol{u}, \boldsymbol{u}) \geq \alpha||\boldsymbol{u}||^2_{H^1(\Omega)}, \forall \boldsymbol{u} \neq 0, \alpha > 0$), however, the bilinear form is symmetric and defines an inner product, called the **energy** inner product. The energy of $\boldsymbol{u}$ is given by $a(\boldsymbol{u}, \boldsymbol{u})$.

To simplify the notation while remaining mathematically rigorous, throughout the rest of this section we will only consider the self-adjoint uniformly elliptic case. The more general case is discussed briefly in Chapter 5.

1.6.1 Projections

The concept of a projection is so important for developing an understanding of domain decomposition and multilevel methods that we give an elementary introduction. Before defining projections we must introduce the objects we are dealing with and the way we measure distances and angles.

Assume we are given a vector space V with an inner product $a(\cdot, \cdot)$. For instance, V may be the usual Euclidean space $\mathcal{R}^N$ and $a(u, v) = u^T Av$, where A is a symmetric, positive definite matrix; that is, all of its eigenvalues are positive. Associated with the inner product $a(\cdot, \cdot)$ is the **a-norm** defined by $||\boldsymbol{u}||_a = \sqrt{a(\boldsymbol{u}, \boldsymbol{u})}$. The norm measures, in some sense, the length of the vector $\boldsymbol{u}$. The norm $||\boldsymbol{u} - \boldsymbol{v}||_a$ measures the distance between two vectors $\boldsymbol{u}$ and $\boldsymbol{v}$. The standard Euclidean inner product is simply $u^T v$, that is, $A = I$, the identity operator. The standard Euclidean norm is given by $||u||_2 = \sqrt{u^T u} = \sqrt{\sum_{i=1}^N u_i^2}$. In analyzing domain decomposition and multilevel methods the important vector spaces are usually function spaces, such as $H_0^1(\Omega)$.

Let V_1 be a subspace of V; we denote this by $V_1 \subset V$. Let $\boldsymbol{e} \in V$ be any element of V. A very natural question is to find the element in V_1 that is "closest" to $\boldsymbol{e}$. That is, which element $\boldsymbol{e}_1 \in V_1$ minimizes the distance between $\boldsymbol{e}$ and $\boldsymbol{e}_1$? We formally define the **projection** of $\boldsymbol{e}$ onto the subspace V_1, in the inner product $a(\cdot, \cdot)$, by

$$\boldsymbol{e}_1 = P\boldsymbol{e} = \arg \inf_{\boldsymbol{v} \in V_1} ||\boldsymbol{e} - \boldsymbol{v}||_a. \tag{1.19}$$

There is a very convenient alternative definition of $\boldsymbol{e}_1 = P\boldsymbol{e}$ given by the following: Find $\boldsymbol{v}_1 \in V_1$ so that

$$a(\boldsymbol{e}_1, \boldsymbol{v}) = a(\boldsymbol{e}, \boldsymbol{v}) \qquad \forall \boldsymbol{v} \in V_1 \tag{1.20}$$

or, equivalently,

$$a(\boldsymbol{e}_1 - \boldsymbol{e}, \boldsymbol{v}) = 0 \qquad \forall \boldsymbol{v} \in V_1.$$

This is another way of saying that $\boldsymbol{e}_1 - \boldsymbol{e}$ is orthogonal, in the $a(\cdot, \cdot)$ inner product, to all elements of V_1.

We now show that the solution of (1.20) is the minimizer of (1.19).

$$\begin{aligned} ||\boldsymbol{e}_1 - \boldsymbol{e}||_a^2 &= a(\boldsymbol{e}_1 - \boldsymbol{e}, \boldsymbol{e}_1 - \boldsymbol{e}), \\ &= a(\boldsymbol{e}_1 - \boldsymbol{e}, \boldsymbol{v} - \boldsymbol{e}) \qquad \forall \boldsymbol{v} \in V_1, \\ &\le ||\boldsymbol{e}_1 - \boldsymbol{e}||_a ||\boldsymbol{v} - \boldsymbol{e}||_a \quad \forall \boldsymbol{v} \in V_1. \end{aligned}$$

Therefore, by dividing through by $||\boldsymbol{e}_1 - \boldsymbol{e}||_a$ we obtain

$$||\boldsymbol{e}_1 - \boldsymbol{e}||_a \le ||\boldsymbol{v} - \boldsymbol{e}||_a \qquad \forall \boldsymbol{v} \in V_1.$$

In the special case that $\boldsymbol{e} \in V_1$ the projection of $\boldsymbol{e}$ is exactly $\boldsymbol{e}$, that is, $\boldsymbol{e}_1 = P\boldsymbol{e} = \boldsymbol{e}$.

When $V = \mathcal{R}^N$ and V_1 is the span of the columns of R^T then the projection may be written as the matrix

$$P = R^T (RAR^T)^{-1} RA.$$

This may be derived from (1.20). Since any element in V_1 is a linear combination of the columns of R^T, $e_1 = R^T \tilde{e}_1$ for some $\tilde{e}_1$. Similarly, $v = R^T \tilde{v}$. Plugging this into (1.20) gives

$$\begin{aligned} \tilde{v}^T RAR^T \tilde{e}_1 &= a(\boldsymbol{e}_1, \boldsymbol{v}), \\ &= a(\boldsymbol{e}, \boldsymbol{v}) \qquad \forall v \in \text{span}\{R^T\}, \\ &= \tilde{v}^T RAe \qquad \forall \tilde{v}. \end{aligned}$$

Thus

$$(RAR^T)\tilde{e}_1 = RAe$$

or

$$\tilde{e}_1 = (RAR^T)^{-1} RAe,$$

which implies

$$Pe = e_1 = R^T \tilde{e}_1 = R^T (RAR^T)^{-1} RAe.$$

This calculation is essentially the same as that given in Subsection 1.1.3 except that it uses the alternative definition of projection.

1.6.2 Matching Grids

We next demonstrate that in the case of matching grids the correction terms generated in one level Schwarz methods can often be viewed as projections of the error. Let $\boldsymbol{u}^n$ denote the finite element solution on the entire domain at the end of the nth iteration, while $\boldsymbol{u}^{n+1/2}$ denotes the finite element solution on the entire domain at the end of the first substep of iteration $n+1$.

We can rewrite (1.2) and (1.3) as

$$\begin{aligned} L(\boldsymbol{u}^{n+1/2} - \boldsymbol{u}^n) &= -L\boldsymbol{u}^n + f && \text{in } \Omega_1, \\ \boldsymbol{u}^{n+1/2} - \boldsymbol{u}^n &= 0 && \text{on } \partial\Omega_1, \\ \boldsymbol{u}^{n+1/2} - \boldsymbol{u}^n &= 0 && \text{in } \Omega_2 \setminus \Omega_1, \end{aligned}$$

and

$$\begin{aligned} L(\boldsymbol{u}^{n+1} - \boldsymbol{u}^{n+1/2}) &= -L\boldsymbol{u}^{n+1/2} + f && \text{in } \Omega_2, \\ \boldsymbol{u}^{n+1} - \boldsymbol{u}^{n+1/2} &= 0 && \text{on } \partial\Omega_2, \\ \boldsymbol{u}^{n+1} - \boldsymbol{u}^{n+1/2} &= 0 && \text{in } \Omega_1 \setminus \Omega_2. \end{aligned}$$

This can be expressed in weak form as

$$a(\boldsymbol{u}^{n+1/2} - \boldsymbol{u}^n, \boldsymbol{v}) = f(\boldsymbol{v}) - a(\boldsymbol{u}^n, \boldsymbol{v}) \qquad \forall \boldsymbol{v} \in H_0^1(\Omega_1) \tag{1.21}$$

and

$$a(\boldsymbol{u}^{n+1} - \boldsymbol{u}^{n+1/2}, \boldsymbol{v}) = f(\boldsymbol{v}) - a(\boldsymbol{u}^{n+1/2}, \boldsymbol{v}) \qquad \forall \boldsymbol{v} \in H_0^1(\Omega_2), \tag{1.22}$$

where $\boldsymbol{u}^{n+1/2} - \boldsymbol{u}^n \in H_0^1(\Omega_1)$ and $\boldsymbol{u}^{n+1} - \boldsymbol{u}^{n+1/2} \in H_0^1(\Omega_2)$.

Let $\boldsymbol{e}^n = \boldsymbol{u}^* - \boldsymbol{u}^n$ be the error at the nth iteration; then

$$a(\boldsymbol{u}^{n+1/2} - \boldsymbol{u}^n, \boldsymbol{v}) = a(\boldsymbol{e}^n, \boldsymbol{v}) \qquad \forall \boldsymbol{v} \in H_0^1(\Omega_1)$$

and

$$a(\boldsymbol{u}^{n+1} - \boldsymbol{u}^{n+1/2}, \boldsymbol{v}) = a(\boldsymbol{e}^{n+1/2}, \boldsymbol{v}) \qquad \forall \boldsymbol{v} \in H_0^1(\Omega_2).$$

We define a projection, $\boldsymbol{P}_i\boldsymbol{u}$, in the inner product $a(\cdot,\cdot)$, by

$$a(\boldsymbol{P}_i\boldsymbol{u}, \boldsymbol{v}) = a(\boldsymbol{u}, \boldsymbol{v}) \qquad \boldsymbol{P}_i\boldsymbol{u} \in H_0^1(\Omega_i),\ \forall \boldsymbol{v} \in H_0^1(\Omega_i).$$

At each half-step the corrections calculated, $\boldsymbol{u}^{n+1/2} - \boldsymbol{u}^n$ and $\boldsymbol{u}^{n+1} - \boldsymbol{u}^{n+1/2}$, are thus the projection of the error onto the subspace $H_0^1(\Omega_1)$ or $H_0^1(\Omega_2)$.

In the finite element method, the space $H_0^1(\Omega)$ is replaced with a finite dimensional subspace $V^h \subset H_0^1(\Omega)$. Also the spaces $H_0^1(\Omega_i)$ are replaced by $V_i^h = H_0^1(\Omega_i) \cap V^h$.

The numerical alternating Schwarz method can be written as

$$a(\boldsymbol{u}^{n+1/2} - \boldsymbol{u}^n, \boldsymbol{v}) = f(\boldsymbol{v}) - a(\boldsymbol{u}^n, \boldsymbol{v}) \qquad \forall \boldsymbol{v} \in V_1^h \tag{1.23}$$

and

$$a(\boldsymbol{u}^{n+1} - \boldsymbol{u}^{n+1/2}, \boldsymbol{v}) = f(\boldsymbol{v}) - a(\boldsymbol{u}^{n+1/2}, \boldsymbol{v}) \qquad \forall \boldsymbol{v} \in V_2^h, \tag{1.24}$$

where $\boldsymbol{u}^{n+1/2} - \boldsymbol{u}^n \in V_1^h$ and $\boldsymbol{u}^{n+1} - \boldsymbol{u}^{n+1/2} \in V_2^h$.

Let $\{\phi_k\}$ be the usual nodal basis functions of V^h and $\{\phi_k^i\}$ be the subset of those functions whose support lies in Ω_i (the support of a function is simply the set of all x values for which the function is nonzero). Express $\boldsymbol{u} = \sum_k u_k \phi_k$ and let A be the **stiffness matrix**, that is, $A_{kj} = a(\phi_j, \phi_k)$. It is possible to derive explicit matrix representations for the operators $\boldsymbol{P}_i$.

Let $\boldsymbol{w} = \boldsymbol{P}_i \boldsymbol{u} = \sum_k w_k \phi_k^i$; then

$$a(\boldsymbol{P}_i \boldsymbol{u}, \boldsymbol{v}) = a(\boldsymbol{u}, \boldsymbol{v}) \qquad \forall \boldsymbol{v} \in V_i^h.$$

This implies

$$a\Big(\sum_k w_k \phi_k^i, \phi_l^i\Big) = a\Big(\sum_k u_k \phi_k, \phi_l^i\Big) \quad \forall \phi_l^i,$$
$$\sum_k w_k a(\phi_k^i, \phi_l^i) = \sum_k u_k a(\phi_k, \phi_l^i) \quad \forall \phi_l^i.$$

Recall that the matrix operator R_i maps the coefficients of u to those associated only with Ω_i. Thus we obtain

$$A_{\Omega_i} w = (R_i A R_i^T) w = R_i A u$$

or

$$w = (R_i A R_i^T)^{-1} R_i A u.$$

Hence the matrix representation of the projection operator $\boldsymbol{P}_i$ is given by

$$P_i = R_i^T (R_i A R_i^T)^{-1} R_i A.$$

Note that we distinguish between the projection operator $\boldsymbol{P}_i$ and its matrix representation P_i.

Now, when approximate solvers are used on the subdomains, this means that (1.21) and (1.22) are replaced by

$$a_1(\boldsymbol{u}^{n+1/2} - \boldsymbol{u}^n, \boldsymbol{v}) = f(\boldsymbol{v}) - a(\boldsymbol{u}^n, \boldsymbol{v}) \qquad \forall \boldsymbol{v} \in V_1^h$$

and

$$a_2(\boldsymbol{u}^{n+1} - \boldsymbol{u}^{n+1/2}, \boldsymbol{v}) = f(\boldsymbol{v}) - a(\boldsymbol{u}^{n+1/2}, \boldsymbol{v}) \qquad \forall \boldsymbol{v} \in V_2^h.$$

The bilinear forms $a_i(\cdot,\cdot)$ approximate $a(\cdot,\cdot)$, in some sense, on the subspaces V_i^h. The projection operator $\boldsymbol{P}_i$ is replaced by a projectionlike operator, $\boldsymbol{T}_i$,

$$a_i(\boldsymbol{T}_i\boldsymbol{u},\boldsymbol{v}) = a(\boldsymbol{u},\boldsymbol{v}) \qquad \boldsymbol{T}_i\boldsymbol{u} \in V_i^h,\ \forall \boldsymbol{v} \in V_i^h.$$

By using the same type of calculations as above, the matrix representation of $\boldsymbol{T}_i$ is given by $T_i = R_i^T(\tilde{A}_i)^{-1}R_iA$, where $(\tilde{A}_i)_{kj} = a_i(\phi_j^i, \phi_k^i)$.

1.6.3 Nonmatching Grids

For the case of nonmatching grids, the formulation is a bit trickier because the solution on a subdomain cannot be expressed as the projection of the error of a finite element solution on the entire domain.

We use the fact that the interpolated values, away from the artificial boundary, do not affect the result and define $\tilde{\boldsymbol{u}}_i^n$ from (1.11). Then (1.23) and (1.24), for the nonmatching grid case, may be written as

$$a(\boldsymbol{u}_1^{n+1} - \tilde{\boldsymbol{u}}_1^n, \boldsymbol{v}) = f(\boldsymbol{v}) - a(\tilde{\boldsymbol{u}}_1^n, \boldsymbol{v}) \qquad \forall \boldsymbol{v} \in V_1^h$$

and

$$a(\boldsymbol{u}_2^{n+1} - \tilde{\boldsymbol{u}}_2^n, \boldsymbol{v}) = f(\boldsymbol{v}) - a(\tilde{\boldsymbol{u}}_2^n, \boldsymbol{v}) \qquad \forall \boldsymbol{v} \in V_2^h.$$

Let $\boldsymbol{u}_i^*$ be the finite element solutions in Ω_i, that is, the functional representation of the solution to the linear system (1.10). Then

$$a(\boldsymbol{u}_1^{n+1} - \tilde{\boldsymbol{u}}_1^n, \boldsymbol{v}) = a(\tilde{\boldsymbol{e}}_1^n, \boldsymbol{v}) \qquad \forall \boldsymbol{v} \in V_1^h$$

and

$$a(\boldsymbol{u}_2^{n+1} - \tilde{\boldsymbol{u}}_2^n, \boldsymbol{v}) = a(\tilde{\boldsymbol{e}}_2^n, \boldsymbol{v}) \qquad \forall \boldsymbol{v} \in V_2^h,$$

where $\tilde{\boldsymbol{e}}_i^n = \boldsymbol{u}_i^* - \tilde{\boldsymbol{u}}_i^n$. So we see that for symmetric $a(\cdot,\cdot)$ the correction term $\boldsymbol{u}_i^{n+1} - \tilde{\boldsymbol{u}}_i^n$ is still a projection of the most recent error onto the finite element subspace V_i^h.

Given these derivations it is now possible to provide a simple interpretation of the alternating Schwarz method for two (or any number of) subdomains. One projects the error onto the first subdomain and removes it. Next the error is projected onto the second subdomain and removed. Since the process of projecting the error onto the second subdomain introduces new error in the first subdomain, we must continue the process until the remaining errors are sufficiently small.

There is a nice simple physical analogy to this process, due to Schwarz himself. Consider two pools of water which overlap and share a common region. Now assume each pool has a separate movable gate and a pump. Furthermore, a pool's pump may only be run when its gate is down. In the first step, one closes the gate for the first pool and pumps out all of the water. Then one opens that pool's gate and the remaining water equalizes between the two pools. Now close the gate for the second pool and pump out its water. Clearly each repetition of this process will further drain the pools.

Notes and References for Section 1.6

- An elementary introduction to the finite element Galerkin method may be found in Johnson [Joh87]. More advanced references are Ciarlet [Cia78] or Strang and Fix [Str73].
- The formulation of the alternating Schwarz method in terms of projections was discussed as early as 1956 by Morgenstern [Mor56].
- The variational formulation has been used by P. Lions, [Lio88].
- The proofs of convergence for Schwarz methods for nonmatching grids, that we are aware of, are based on a discrete maximum principle; see, for example, Miller [Mil65] and Starius [Sta77].

2

Two Level Algorithms

AS WE HAVE SEEN in the previous chapter, single level methods are most effective only for a small number of subdomains. The problem is intrinsic in the mathematics (and the physics!) of elliptic (diffusion) dominated PDEs. This problem can be understood in several different ways; we first give a simple linear algebra explanation of why using only local solvers is not completely satisfactory for efficient numerical solvers of elliptic dominated PDEs.

2.1 Subdomain Solves Are Not Sufficient

Consider an error as depicted in Figure 2.1; the error is constant over a region, $\tilde{\Omega}_i$ that includes Ω_i. The local correction is given by

$$c_i = \begin{pmatrix} 0 & & \\ & A_{\Omega_i}^{-1} & \\ & & 0 \end{pmatrix} (f - Au^n) = \begin{pmatrix} 0 & & \\ & A_{\Omega_i}^{-1} & \\ & & 0 \end{pmatrix} Ae^n.$$

For the standard finite difference and finite element discretizations of second order elliptic PDEs the row sum of A corresponding to a point away from any boundary is zero (this is because A corresponds to a differential operator with no zero-order term). Now since e^n is constant on $\tilde{\Omega}_i$ the vector Ae^n is exactly zero on Ω_i. Hence the correction term c_i is also zero.

In the language of Section 1.6 the projection of a constant error onto the subspace V_i^h is zero. This is because the space V_i^h does not contain the constant functions, since all functions in V_i^h are zero on the boundary. (Alas, using Neumann or Robin boundary conditions on the artificial boundaries does not cure this problem, since it is not possible to determine locally the constant shift that should be included in the correction.) For subdomains adjacent to a physical boundary with Dirichlet boundary conditions this problem does not occur since the error is zero along the boundary. (For Neumann boundaries the problem persists even along the boundary.)

For smooth errors, in a small subdomain, the error may be written as $\epsilon = \bar{\epsilon} + \epsilon_{small}$, where $\bar{\epsilon}$ is the average of ϵ on the subdomain. Since no reduction in the $\bar{\epsilon}$ term may be obtained through a local solve, one can expect little reduction in the entire error ϵ over that subdomain. So we see that for smooth errors the local corrections are ineffectual except near the Dirichlet boundary. Hence we need another inexpensive and easy way of correcting smooth errors.

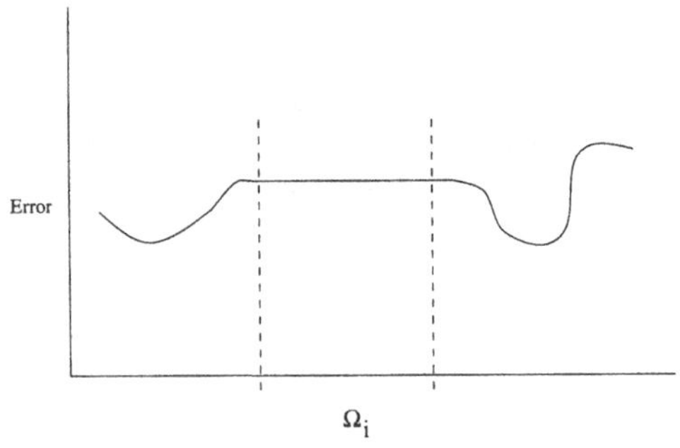

Figure 2.1. Constant Error on Ω_i

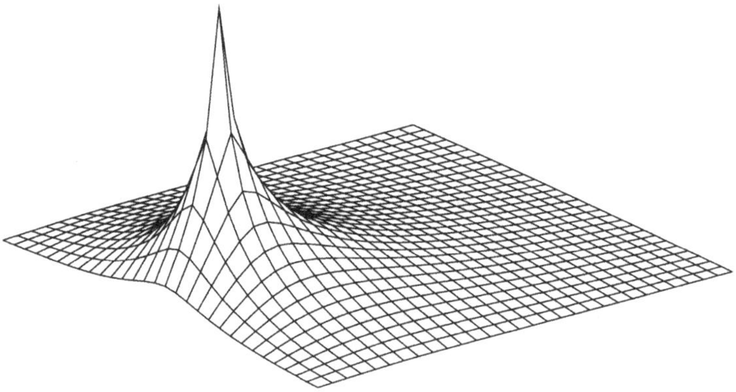

Figure 2.2. Decay of Green's Function

For elliptic PDEs, the solution at any point depends on the right hand side and boundary conditions for the entire domain. We briefly review this dependence for linear problems to provide motivation for the two and multilevel algorithms.

Again, for simplicity, we consider the model problem

$$\begin{aligned} \boldsymbol{Lu} &= \boldsymbol{f} \quad \text{in } \Omega, \\ \boldsymbol{u} &= 0 \quad \text{on } \partial\Omega. \end{aligned}$$

The explicit solution can be written as

$$\boldsymbol{u}(x) = \int G(x, x')\boldsymbol{f}(x')dx',$$

where $G(x, x')$ is the appropriate Green's function. In Figure 2.2 we plot $G(x_0, x')$ as a function of x' for a fixed x_0, in a square region. The function has a characteristic peak around the value $x' = x_0$. We see that the solution at x_0 therefore depends strongly on the value of $\boldsymbol{f}(x')$ for x' near x_0 and less strongly for x' far from x_0. This behavior is characteristic of all elliptic PDEs. Algorithms for the numerical solution of elliptic PDEs should reflect this fact. More computational work should be devoted to calculating the effect of the right hand side that is mostly local, but without ignoring the

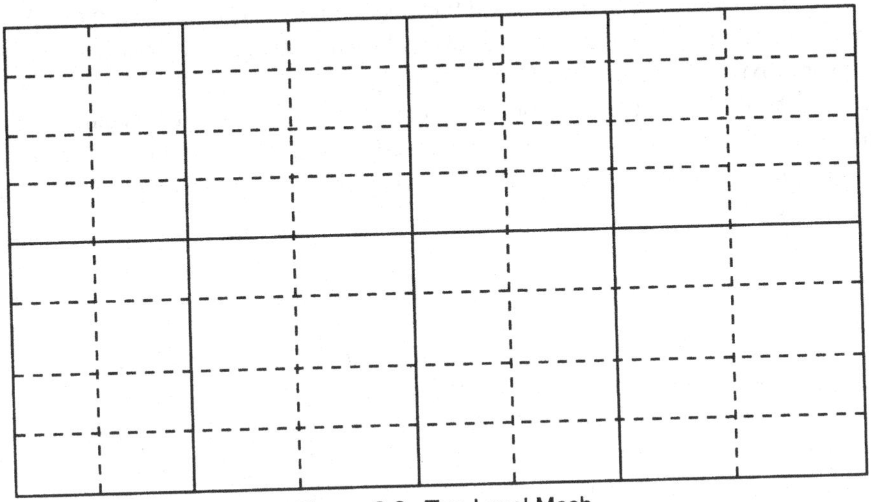

Figure 2.3. Two Level Mesh

long range influence. Note also that this formula is very similar to the matrix formula

$$u = A^{-1}f,$$

in fact, $G(x, x')$ may be thought of as a continuous version of A^{-1}.

The problem with single level methods is that the information about the right hand side at one point is conveyed through to another point *only* by passing through all the intermediate subdomains, each of which muddles it up a bit more.

The solution to this dilemma has been known for quite a while. The iterative linear system solver must have some mechanism for the global communication of information at each iteration. The best known of these techniques are the **multigrid** methods.

Notes and References for Section 2.1

- The concept of smooth errors actually depends on the operator L. In fact, for a general L the intuitive idea of smooth functions is not correct. One really should say *low energy* functions in V^h as measured in the bilinear form, $a(\cdot, \cdot)$, induced by L. For the Laplacian, the low energy functions are exactly the intuitive "smooth" functions. See Chapter 5 for a brief discussion of the mathematical issues related to this.

2.2 A Simple Two Level Method

In this section we define and derive a simple two level preconditioner (iterative method). Consider the mesh depicted in Figure 2.3, and assume that we have discretized a linear PDE on this domain by using the fine mesh and have obtained the linear system,

$$A_F u_F = f.$$

If we knew the error on the coarser grid, then we could linearly interpolate it back to the fine grid and use this as a correction. The error e_C on the coarse grid is, of course, unknown; but it could be calculated from the error equation $A_C e_C = r_C$ if the residual r_C on the coarse grid were known. The residual on the fine grid is known, so we can use it to approximate the residual on the coarse grid.

Let R^T be the matrix representation of linear **interpolation** from the coarse grid to the fine grid. The operator R is called the **restriction** (matrix) operator. Let A_C be a discrete form of the operator on the coarse grid. A simple **coarse grid correction** could be given by

$$u_F \leftarrow u_F + R^T A_C^{-1} R(f - A_F u_F).$$

That is, we calculate the residual on the fine grid, restrict it to the coarse grid, solve the coarse grid problem, and interpolate the coarse grid solution to the fine grid.

It is not possible, however, simply to use $R^T A_C^{-1} R$ alone as the preconditioner, since it has a large null space (because the rank of $R^T A_C^{-1} R$ is equal to the dimension of A_C, which is much less than the dimension of A_F). Any component of $f - A_F u_F$ which lies in the null space of $R^T A_C^{-1} R$ would never be corrected. In particular, those error components that are not corrected include most of the high frequency error. We call the term $R^T A_C^{-1} R$ the **coarse grid part of the preconditioner**. Thus, we define $B_C = R^T A_C^{-1} R$ and introduce a complete preconditioner by also defining a preconditioner B_F. We can take B_F to be any preconditioner for the matrix A_F having full rank. A simple example would be to take $B_F = D_F^{-1}$, the inverse of the diagonal of A_F. In general, B_F will be **localized** (and therefore easier to use in a parallel algorithm and better suited for memory hierarchies), reflecting the local properties of the operator, while the B_C will be designed to represent the long range effects of the elliptic PDE. The choice $B_F = D_F^{-1}$ is the extreme case of pointwise local solves. More often in domain decomposition algorithms the application of B_F would involve subdomain solves, corresponding to the preconditioners discussed in Chapter 1.

We can now introduce a two step preconditioner which contains both short range (B_F) and long range (B_C) components

$$\begin{aligned} u_F^{n+1/2} &\leftarrow u_F^n + B_C(f - A_F u_F^n), \\ u_F^{n+1} &\leftarrow u_F^{n+1/2} + B_F(f - A_F u_F^{n+1/2}). \end{aligned} \tag{2.1}$$

In the special case when $B_F = D_F^{-1}$, the second half-step is simply one step of the Richardson method with Jacobi preconditioning. This algorithm contains the essential features of two level algorithms: a coarse grid correction and local solvers used in combination as a preconditioner. Most often, the B_F part of the preconditioner does a poor job of correcting the low frequency (smooth) components of the error, but these are the modes which are well resolved by the coarse grid correction.

As always, (2.1) can be written as a one step method,

$$u_F^{n+1} \leftarrow u_F^n + (B_C + B_F - B_F A_F B_C)(f - A_F u_F^n).$$

Note that this preconditioner has the exact same structure as (1.14) in Chapter 1, where we introduced a preconditioner consisting of two (local) parts. The preconditioner can

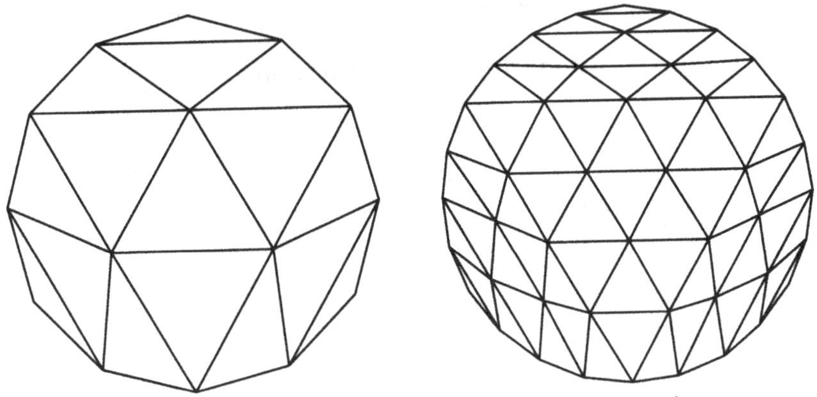

Figure 2.4. Nonnested Grids due to Curved Boundaries

be written in the form

$$B = B_C + B_F - B_F A_F B_C.$$

The application of B to a vector can be calculated by one iteration of (2.1) with a *zero* initial guess, just as in the alternating Schwarz method; that is, $v = Br$ is given by

$$v \leftarrow B_C r,$$
$$v \leftarrow v + B_F(r - A_F v).$$

There is no intrinsic need for the two grids to be nested. The algorithms discussed in this chapter may be used directly with nonnested grids as depicted in Figure 2.4.

Notes and References for Section 2.2

- For nonsymmetric problems there is no algorithmic need for the interpolation operator to be the transpose of the restriction operator R. The coarse part of our preconditioner B_C may therefore take the more general form $B_C = \mathcal{I} A_C^{-1} R$, where $\mathcal{I}$ is the interpolation operator and R is the restriction operator. However, in practice, and for theoretical reasons, it is often best for the interpolation matrix to be the transpose of the restriction matrix. See Section 2.8.
- The first two level method appears to have been suggested by Southwell [Sou35a] and [Sou35b].

2.3 General Two Level Methods

In this section we generalize the method introduced above and show how a multitude of algorithms may be designed by using a few basic principles.

The general **multiplicative two level method** is given by

$$u_F^{n+1/2} \leftarrow u_F^n + B_C(f - A_F u_F^n),$$
$$u_F^{n+1} \leftarrow u_F^{n+1/2} + B_F(f - A_F u_F^{n+1/2}),$$

where B_F is any preconditioner.

Algorithm 2.3.1: Smoothing

- $u \leftarrow 0$.
- For $n = 1, \ldots, n_{\text{step}}$

  $u \leftarrow u + \sum_i R_i^T A_{\Omega_i}^{-1} R_i (r - A_F u)$
- EndFor.

In classical multigrid, the local solvers, B_F, are usually a simple iterative scheme such as Jacobi or Gauss-Seidel. They are referred to as **smoothers** because they remove the high frequency components of the error. It is possible to view domain decomposition as a generalization of multigrid where the simple iterative smoothers are allowed to be more general and more robust.

As with the alternating Schwarz method, the multiplicative two level methods have corresponding additive forms, the **additive two level method,**

$$u_F^{n+1} \leftarrow u_F^n + (B_C + B_F)(f - A_F u_F^n).$$

Again, this would not be used directly, but as part of a Krylov subspace accelerator (see Appendix 1.2).

The coarse grid correction only resolves the part of the error which can be represented on the coarse grid, that is, the components of the error that are smooth. It cannot resolve the part of the error with high frequency components. Therefore the preconditioner B_F should, at least, do a good job of correcting the high frequency components of the error.

In multigrid language, we are including the use of **Schwarz smoothers.** Schwarz smoothers are simply one level additive or multiplicative overlapping Schwarz preconditioners. For instance, the additive Schwarz smoother applied to the residual r is given by

$$u \leftarrow \sum_i R_i^T A_{\Omega_i}^{-1} R_i r.$$

Jacobi and Gauss-Seidel smoothers correspond to the special case where the local restriction operators R_k return a single component of the vector. Block Jacobi and block Gauss-Seidel are the special cases in which there is no overlap between subdomains. Two or more smoothing steps may also be used, for instance, as in Algorithm 2.3.1.

In **two level Schwarz** methods the local preconditioner, B_F, is either a multiplicative or an additive overlapping Schwarz preconditioner (i.e., a Schwarz smoother). These can be combined in either an additive or multiplicative two level method, so one obtains several distinct preconditioners (Algorithms 2.3.2 and 2.3.3). For notational convenience when the fine grid component of our preconditioner B_F is constructed in this manner (from $B_i = R_i^T A_{\Omega_i}^{-1} R_i$, $i = 1, 2, \ldots, p$) we will reserve the index zero for the coarse part of the preconditioner; that is, $B_0 \equiv B_C = R^T A_C^{-1} R$.

Algorithm 2.3.2: Two Level Additive Schwarz Preconditioner

$$v \leftarrow \left(R^T A_C^{-1} R + \sum_{i=1}^{p} B_i\right) r.$$

Algorithm 2.3.3: Two Level Multiplicative Schwarz Preconditioner

$$
\begin{aligned}
v &\leftarrow R^T A_C^{-1} Rr \\
v &\leftarrow v + B_1(r - A_F v) \\
&\cdots \\
v &\leftarrow v + B_p(r - A_F v).
\end{aligned}
$$

Algorithm 2.3.4: Two Level Hybrid I Schwarz Preconditioner

$$
\begin{aligned}
v &\leftarrow B_1 r \\
v &\leftarrow v + B_2(r - A_F v) \\
&\cdots \\
v &\leftarrow v + B_p(r - A_F v) \\
v &\leftarrow v + R^T A_C^{-1} Rr.
\end{aligned}
$$

For Algorithm 2.3.2 the preconditioner may be written explicitly as

$$B = \sum_{i=0}^{p} B_i = R^T A_C^{-1} R + \sum_{i=1}^{p} B_i.$$

For the multiplicative form, Algorithm 2.3.3, the preconditioner is given by

$$B = [I - (I - B_p A_F) \cdots (I - B_1 A_F)(I - B_0 A_F)] A_F^{-1};$$

for Algorithm 2.3.4, it is given by

$$B = B_0 + [I - (I - B_p A_F) \cdots (I - B_1 A_F)] A_F^{-1};$$

Algorithm 2.3.5: Two Level Hybrid II Schwarz Preconditioner

$$v \leftarrow \sum_{i=1}^{p} B_i r$$

$$v \leftarrow v + R^T A_C^{-1} R(r - A_F v).$$

and for Algorithm 2.3.5, it is given by

$$B = B_0 + \left(I - B_0 A_F\right)\left(\sum_{i=1}^{p} B_i\right).$$

Various other combinations are possible; for instance, one could do two sweeps of a one level multiplicative Schwarz method for each coarse grid correction. In addition, the multiplicative, hybrid I and hybrid II methods may all be symmetrized, by sweeping back through the subdomains. As we have seen and will see in succeeding chapters, all of these methods have their own advantages and disadvantages in terms of parallelism, convergence rate, and so forth.

Notes and References for Section 2.3

- The two level overlapping additive Schwarz method is due to Dryja and Widlund [Dry87], [Dry89].
- The two level hybrid I overlapping Schwarz method was proposed and analyzed by X. Cai [Cai93].
- Mandel introduced the two level hybrid II Schwarz method in a slightly different context in [Man93]; see also Section 4.3.3.

2.4 Coarse Grid Corrections

There are virtually unlimited choices of the coarse grid correction that may be used. Convergence of the entire scheme will depend strongly on the particular interpolation and coarse grid operator used. The type of interpolation, R_0^T, and coarse grid operator, A_C, should not be viewed as two independent choices, since the convergence rate depends on the interaction of the two. A rule of thumb is that once one of the two has been chosen, the other one should be picked so that the coarse grid correction is as close as possible to being a projection of the error (onto a coarse subspace); see Section 2.8.

Recall from Chapter 1 that the form of the local preconditioning matrix is given by $B_i = R_i^T A_{\Omega_i}^{-1} R_i$, where $A_{\Omega_i} = R_i A R_i^T$. In Sections 1.1.3 and 1.6, we demonstrated that the local correction $c_i = B_i r$ was a projection of the error onto the subspace spanned by the columns of R_i^T. A **Galerkin or variational coarse grid correction** has the property that $A_C = R_0 A R_0^T$. Methods that use Galerkin coarse grid corrections are often called **nested** two level (or multilevel) methods. The reason that Galerkin coarse grid corrections are interesting is that the term $B_C A_F = R_0^T A_C^{-1} R_0 A_F$ is also a projection; compare this to (1.17). In fact, a Galerkin coarse grid correction is a projection of the error onto a "coarse" subspace given by $V_0 = \text{span}\{R_0^T\}$; see Section 2.8, where this is demonstrated in a finite element context.

When using conforming finite elements, the selection of the coarse grid correction is usually simple. On the coarse triangulation one simply uses a finite element space of degree less than or equal to that used for the fine space. The natural interpolation and coarse grid operator can then be derived immediately and a Galerkin coarse grid correction is automatically obtained.

2.5 Convergence Behavior

For symmetric problems the convergence behavior can be quite good, requiring on the order of 10 iterations (for a reduction in the norm of the error of 10^{-5}), independent of the problem size. For nonsymmetric problems, if H is small enough (depending on the strength of the convection term; see Section 5.4.2), the same satisfactory behavior is retained. More specifically, referring to Figure 1.10 we have for symmetric problems, both theoretically and numerically:

- if δ is kept proportional to H, the number of iterations is bounded independently of h, H, and H/h,
- the number of iterations for the multiplicative Schwarz method is roughly half of that needed for the additive Schwarz method, and
- convergence is poor for $\delta = 0$ but improves rapidly as the overlap δ is increased.

As with the one level methods (see Section 1.4), using a small overlap between the local subdomains improves the convergence. However, the improvement is less dramatic than when no coarse grid solver is used, because the underlying convergence rate is already so much better as a result of the use of the coarse grid correction.

These results are quite remarkable. With the addition of a (possibly quite small) coarse grid problem, the rate of convergence has been made independent of the size of the problem. This behavior reflects the fact that the Green's function for the solution of the PDE, while nonzero everywhere within the domain, is relatively small in magnitude away from its peak. The local subdomains, being solved to full accuracy, capture the behavior near the peak of the Green's function. The coarse grid solve captures the behavior of the tail, which, because it is small in magnitude, can be approximated by a system with fewer degrees of freedom.

Computational Results 2.5.1: Studies with Many Subdomains

Purpose: Demonstrate the convergence behavior of additive and multiplicative Schwarz methods for many subdomains with a coarse grid correction.

PDE: The Poisson equation,

$$-\Delta \boldsymbol{u} = xe^y \quad \text{in } \Omega,$$
$$\boldsymbol{u} = -xe^y \quad \text{on } \partial\Omega,$$

with Dirichlet boundary conditions.

Domain: Unit square, with mesh spacing $h = 1/(N+1)$.

Discretization: Piecewise linear finite elements.

Calculations: For this problem, we consider a fixed number of subdomains with variable overlap as well as a fixed overlap and a variable number of subdomains. The Krylov subspace method used is GMRES with a restart of 10.

Iteration Counts for a 10^{-2} Reduction in Error

		Multiplicative with Overlap			Additive with Overlap		
N	Decomposition	1	2	4	1	2	4
33	2,2	4	4	3	5	5	5
33	4,4	4	4	3	6	5	5
33	8,8	4	3	2	4	5	7
33	16,16	3	2	2	9	9	9
65	2,2	6	5	4	6	6	6
65	4,4	4	4	5	6	5	5
65	8,8	4	4	4	7	5	6
65	16,16	4	3	2	5	7	10

Discussion: Note the insensitivity of the iteration count as a function of the number of domains. Contrast this with the tables for Computational Results 1.4.1, where much more modest increases in the number of subdomains significantly increased the iteration count.

An interesting feature with these examples is that the additive version, while taking more iterations than the multiplicative, requires fewer than twice as many iterations. An exception is for many subdomains, for example, for the (16, 16) decomposition, where the error reflects the concentration of the remaining residual along the joints between the subdomains in the additive method. Another interesting feature is the behavior of the additive method when the coarse problem is very fine. In this case, adding more overlap actually increases the iteration count! Note that, in (16, 16) decomposition for a grid with size 65×65, the decomposition divides the grid into subgrids that are 5×5. An overlap of 4 extends this subdomain completely over the adjacent subdomains. An overlap of 2 extends halfway across the subdomains, where the edge of the overlap will meet the edge of the overlap from another subdomain. In both of these rather extreme cases, the additive method behaves poorly.

Computational Results 2.5.2: Studies with Many Subdomains

Purpose: Demonstrate the effect of the convection term on the efficiency of the coarse grid correction.

PDE: Convection-diffusion

$$-\Delta u + \beta \cdot \nabla u = xe^y \quad \text{in } \Omega,$$
$$u = -xe^y \quad \text{on } \partial\Omega,$$

with Dirichlet boundary conditions. The value β is a vector that gives the direction of convection; in the calculations below, the angle in the descriptions is the angle that this vector makes with the x-axis.

Domain: Unit square.

Discretization: Centered finite differences with upwinding.

Calculations:

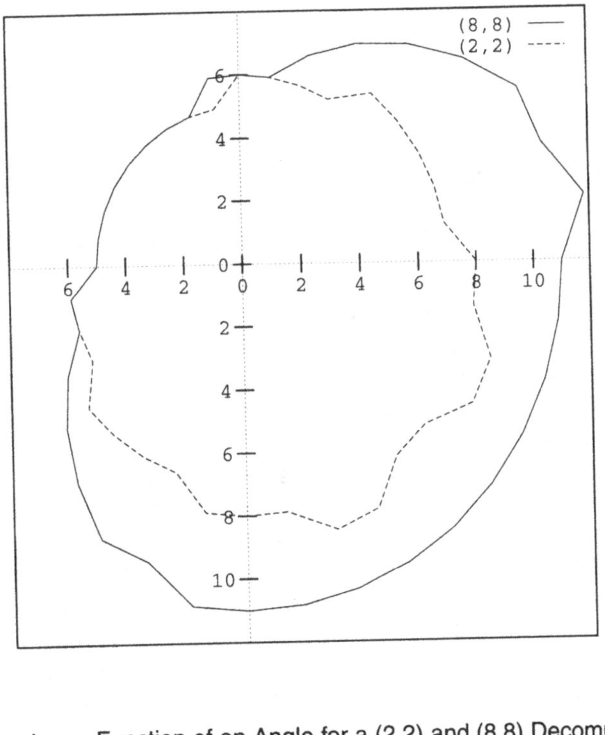

Plot of Iteration Count as a Function of an Angle for a (2,2) and (8,8) Decomposition with an Overlap of 1, a 65 × 65 Mesh, and the Multiplicative Schwarz Method. The Magnitude of β is 10.

The Krylov subspace method used is GMRES with a restart of 10 for both graphs.

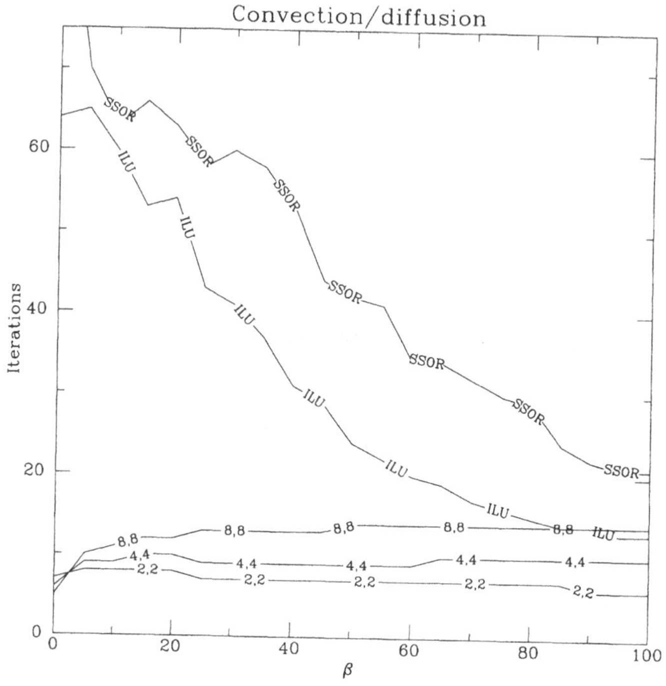

Plot of Iteration Count as a Function of β for an Angle of Zero Degrees, for a (2,2), (4,4), and (8,8) Decomposition with an Overlap of 1, a 65×65 Mesh, and Multiplicative Schwarz Method. For Comparison, ILU(0) and SSOR Preconditioned GMRES (on a Single Domain) Are Also Shown.

Discussion: These two graphs show the effects of the convection term on a two level domain decomposition method. The first graph compares a (2,2) with an (8,8) decomposition for a fairly large value of β, the convection parameter. The performance is generally quite good (iteration count is low) though somewhat sensitive to the angle.

The second graph takes a fixed angle and compares the performance of domain decomposition with ILU and SSOR as a function of the convection parameter. This graph has several interesting features. Most notable is that, as β increases, the performance of both ILU and SSOR improve; for large β, they converge faster than domain decomposition. In essence, when the convection term in the PDE becomes dominant, the best numerical method is the method of characteristics, and (particularly for the angle chosen in this example) both SSOR and ILU are emulating that method (in the limit as $\beta \to \infty$, outside boundary layers, only a single step of SSOR or ILU should be required for convergence when the grid is aligned with the convection).

There is another notable feature to the second graph. Note that for small β, the fastest (in tems of fewest iterations) method is the (8,8) decomposition, but that as β increases, it quickly becomes a poorer method than the (2,2) decomposition. This is consistent with the convergence results that require that the coarse grid

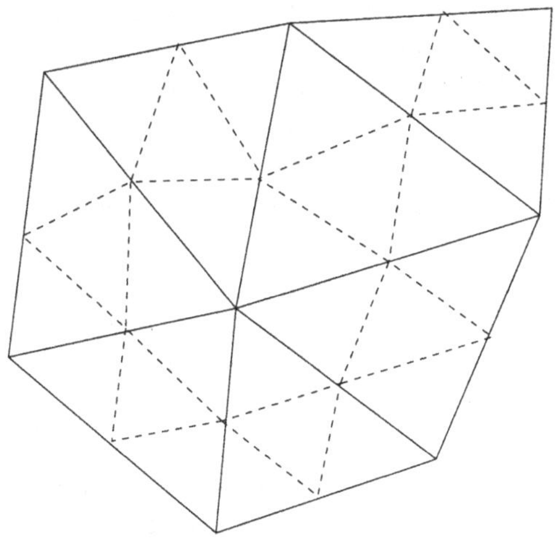

Figure 2.5. Uniform Refinement

be suitably fine, see Section 5.4, where that criterion is relative to the strength of the convection term (i.e., β).

Finally, note that the performance of the two level domain decomposition algorithm is roughly independent of the strength of the convection, suggesting that domain decomposition may be more robust than other preconditioners in complex situations.

Note that this is a model problem and is not intended as an approximation to fluid flow.

2.6 Implementation Issues

In this section we discuss some of the issues of implementation that come up in two level methods. These include the generation of nested grids for the fine and coarse grid problems and the solution of the coarse grid problem itself. Addressing the latter problem provides us with a good opportunity to discuss some issues in parallel programming.

2.6.1 Generation of the Grids

In many applications where the programmer has complete control of the software project, generation of a (nested) coarse grid and coarse grid problem is straightforward. One simply begins with an appropriate finite element or finite difference coarse grid and refines it one or several times to obtain the fine grid. See Figure 2.5, where a triangular mesh is refined (uniformly) by dividing each triangle into four subtriangles. Note that the finer mesh will automatically inherit the shape regularity of the coarser mesh. When the coarse grid is defined in this manner, calculation of the interpolation/restriction is straightforward and inexpensive. For example, consider the calculation of the interpolant of a function on the coarse grid onto the fine grid. Since the fine grid nodes have been obtained from the coarser grid, we know exactly the coarse grid nodes that should be used in interpolating to the fine grid. In addition, the calculations are localized and can be performed in parallel. Refinement of tetrahedra in three dimensions is not so straightforward; however, once a refinement strategy has been chosen, determining the interpolation operators remains simple.

Except for the simplest problems on tensor product grids, we recommend forming the interpolation operators explicitly as sparse matrices. Then each time an interpolation is required this becomes a call to a sparse matrix-vector multiply routine. Similarly, the calculation of $R_0 v$ should be coded as the matrix-vector multiplication of the transpose of the same sparse matrix. (Note that multiplication with the transpose can be done with the same data structures, often by replacing inner product operations with daxpy operations. Daxpy operations are vector operations of the form $y_i = y_i + \alpha x_i$.) In the parallel case one may think of having either a single parallel sparse matrix or individual sparse matrices, one for each processor that applies the interpolation/restriction for the subdomains on that processor. Mathematically these two approaches are identical. The application code in the two cases may be quite different, however. We prefer to conceptualize the process with a single, parallel sparse matrix. In this way we can get more code reuse and do not have to write (and debug!) special purpose code. With a careful implementation, one will not see much performance difference between the two cases on many machines. However, since the parallel sparse matrix-vector operation will get much more code reuse in other applications (and also may be needed for the Krylov subspace code), it will be worth the programmer's time to optimize it. A special purpose code may not be worth the added effort to get top efficiency. (But if a special-purpose code can clearly be shown to be necessary, for example, after performing careful code profiling and measurement, then it is easy to replace the general sparse-matrix operation with a custom implementation.)

When one does not have complete control over the problem generation, it may not be possible or practical to generate the fine grid from the coarse grid in a nested process. Several other approaches are sometimes possible. The simplest may be to run the grid generating procedure that generated the fine grid, again to generate a coarser (not necessarily nested) grid. The construction of the interpolation operator then becomes more difficult. When the two grids are completely unrelated it is necessary to determine the location of all the fine grid points in the coarse grid. When using a naive algorithm this is computationally expensive. On the other hand, if one is solving a time dependent problem and needs to solve elliptic PDEs on the fine grid thousands of times, the initial overhead will pay off since the convergence rate of the linear iterative procedure may be improved significantly by introducing the coarser grids. More sophisticated search techniques may also be employed to reduce the complexity of setting up the interpolation operator.

Another approach is to coarsen the given fine grid. The first step involves calculating a maximal independent set of nodes for the graph of the fine grid. An independent set of nodes in a graph is simply a set of nodes which do not share a common edge. A maximal independent set is simply an independent set for which adding any element will result in a dependent set. One then uses the nodes of the maximal independent set as the coarse grid nodes. The triangulation is then obtained by applying a standard triangulation technique to triangulate the given domain with the given nodes. In triangulating the coarse mesh it is desirable that the resulting triangulation be shape regular; however, in our numerical studies we have observed that if the coarser meshes have a few triangles with very bad aspect ratio, convergence of the linear iterative method is not unduly affected.

Once the coarse grid exists, it is also necessary to have a representation, A_C, of the operator on that grid. For linear problems the easiest way to form A_C is often simply to call the same code that discretized the PDE on the fine grid, but this time with the coarser grid. When this is not possible, one may generate A_C by the sparse product $A_C = R_0 A R_0^T$.

2.6.2 Solution of the Coarse Grid Problem in Parallel

Solving the linear system associated with the coarse grid on a parallel computer presents a potential performance problem. To understand the source of this problem, we will take a brief detour into some elements of time-complexity analysis for parallel computers. We will revisit this discussion in more detail in Chapter 3, where we perform a more careful and complete analysis of the overall computational complexity of domain decomposition methods.

Parallel computers have not only multiple processors but also multiple memories. Many systems, particularly those with large numbers of processors, have memories that are attached to each processor (called local memory to that processor or just local memory); a processor has fast access only to its own local memory.[1] Parallel machines of this type are called distributed memory parallel computers. Most of the larger parallel machines are of this type.

In distributed memory machines, accessing another processor's local memory (e.g., to get data for the overlap into another domain) requires sending some sort of message (which may be a remote memory copy operation). This operation can be modeled as taking time

$$s + rn,$$

where s is the **startup time** or **latency**, n is the number of bytes being transferred, and r is the incremental time to send 1 byte. Table 2.1 shows some sample values for these parameters, as well as f, the time to perform a floating point operation. The MPP is a commercially available distributed memory machine. The "workstation cluster" represents a cluster of fast (and relatively expensive) workstations connected by Ethernet. The "commodity workstation cluster" represents a cluster of low-cost workstations, such as are commonly provided on desktops. Note that s is much larger than the time either to send data or to perform a floating-point operation; this is a common feature of distributed memory systems even when remote memory operations are available. Also, in a workstation cluster, it is important to remember that the bisection bandwidth is much lower than for MPPs; this can reduce the effective bandwidth.

[1] In some cases, the local memory is a large cache.

Table 2.1. *Sample Values for Communication Model*

System	s μsec	r μsec/word	f μsec/flop
MPP	46	0.24	0.008
Workstation cluster (IP)	950	7	0.008
Commodity workstation cluster (IP)	1200	7	0.1

Because of the large cost (relative to floating-point operations) to transfer data from one processor/memory to another, solving small problems in parallel on distributed memory processors can often lead to a slower solution than if only a single processor is used. Since the coarse grid problem is often relatively small, the decision about how to solve it can have a large effect on the parallel performance of the algorithm.

In the case when A_C is quite small, one is faced with two choices: redundantly solve the coarse grid problem on all processors or solve the coarse grid problem on one processor and distribute the results to the other processors. The correct choice between the two approaches depends on the relative speed of computation to communication and whether the other processors can do useful work while one processor is solving the coarse grid problem. In the additive and hybrid I method, the other processors may simultaneously work on their local solves while one processor handles the coarse grid solve. This is not possible for the multiplicative and hybrid II method. Since the multiplicative procedure has the best numerical convergence rate, the decision depends on whether the increased machine utilization outweighs the slower numerical convergence rate of the additive schemes.

For two level algorithms the coarse grid problem A_C is often itself too large to be solved on a single processor. Then one is faced with several choices: solve it by using all the processors working together, solve it by using a subcluster of the processors, or redundantly solve it on groups of processors. Another idea is to calculate the rows of A_C^{-1} explicitly (possibly in parallel) during the setup phase and store them on the appropriate processor. Thus an application of A_C^{-1} can be performed by scattering all the coarse grid nodal values to each processor and having each independently form its portion of the result. In this way only one phase of communication is needed rather than requiring one before the solution of the coarse problem and then a distribution after.

Notes and References for Section 2.6

- The calculation of A_C by explicitly forming RAR^T is sometimes referred to as algebraic multigrid.
- Chan and Smith [Cha95b] performed several numerical studies with coarse grids generated automatically from the fine grid using the techniques outlined above.
- An alternative approach to generating the coarse grid problem and interpolation/restriction operators is given in Bank and Xu [Ban95].
- Additional examples of complexity analysis for other parallel computing issues may be found in [Gro90], [Gro88], and [Fos92].

- The idea of explicitly forming the rows of A_C^{-1} and distributing them among the processors appears to be due to Gropp and Keyes [Gro92a] and independently Fischer [Fis94]. The idea of computing the inverse of the matrix and using matrix multiply with the inverse on parallel machines as a more parallel method for solving linear systems has been suggested by several people.
- See Gropp [Gro92] for a discussion of how to choose where the coarse grid problem should be solved.
- Some discussion of the best size for the coarse grid problem may be found in Chan [Cha95a] and Chan and Shao [Cha94c].

2.7 Fourier Analysis of Two Level Methods

For model elliptic PDEs, on simple domains, discretized by using standard, low order techniques, it is sometimes possible to calculate explicitly the eigenvalues and eigenvectors of the stiffness matrices. In addition, the eigenvalues of the corresponding error propagation matrices for simple iterative methods may be calculated. This is useful for two reasons: it can give explicit bounds for convergence behavior for the model problems and, more importantly, it can indicate the type of behavior one may expect for more realistic PDE models.

To illustrate this we consider the model problem in one dimension,

$$\begin{aligned} -u_{xx} &= f \quad \text{in } (0,1), \\ u(0) &= u(1) = 0. \end{aligned}$$

After this equation has been discretized with centered finite differences (or piecewise linear finite elements) on a uniform grid containing $n = 2N + 1$ interior nodes, one obtains the linear system,

$$A_h u = \frac{1}{h^2} \begin{pmatrix} 2 & -1 & & \\ -1 & 2 & -1 & \\ & \cdots & & \\ & \cdots & -1 & 2 \end{pmatrix} u = f.$$

The mesh width h is given by $1/(n+1)$. The eigenvectors of A_h are given by

$$e^i = \sqrt{2h} \begin{pmatrix} \sin(i\pi h) \\ \sin(2i\pi h) \\ \cdots \\ \sin(ni\pi h) \end{pmatrix},$$

with corresponding eigenvalues

$$\frac{4}{h} \sin^2\left(\frac{i\pi h}{2}\right) = \frac{2}{h}(1 - \cos(i\pi h)) \qquad i = 1, 2, \ldots, n.$$

Observe that the condition number of A_h is given by

$$\begin{aligned}\kappa(A_h) &= \frac{\lambda_{\max}(A_h)}{\lambda_{\min}(A_h)},\\ &= \frac{\sin^2(\frac{n\pi h}{2})}{\sin^2(\frac{\pi h}{2})},\\ &\approx \frac{4}{\pi^2 h^2}.\end{aligned}$$

The error contraction matrix (i.e., $e^{n+1} = E_s e^n$), for Jacobi (diagonal) smoothing, is

$$E_s = I - \frac{h}{2} A_h.$$

Its eigenvalues are given by

$$\begin{aligned}\lambda_i(E_s) &= 1 - \frac{h}{2}\frac{2}{h}[1 - \cos(i\pi h)],\\ &= 1 - [1 - \cos(i\pi h)],\\ &= \cos(i\pi h).\end{aligned}$$

So we see that the reduction in the error depends on the frequency. Error with high frequency, $n/4 < i < 3n/4$, is reduced rapidly while low frequency error is reduced slowly. The error with very high frequency, $i > 3n/4$, is also reduced slowly.

The error contraction factor, given by the spectral radius of E_s, can be estimated by

$$\begin{aligned}\rho(E_s) &= \lambda_{\max}(E_s),\\ &= \cos(\pi h),\\ &\approx 1 - \frac{\pi^2 h^2}{2}.\end{aligned}$$

To analyze the coarse grid correction, we introduce a coarse mesh with $N = (n-1)/2$ interior grid points and a mesh width $H = 1/(N+1) = 2h$. The corresponding stiffness matrix is given by

$$A_H = \frac{1}{H^2}\begin{pmatrix} 2 & -1 & & \\ -1 & 2 & -1 & \\ & \cdots & & \\ & \cdots & -1 & 2 \end{pmatrix}.$$

We consider only piecewise linear restriction and interpolation matrices, that is,

$$R = \begin{pmatrix} 1/2 & 1 & 1/2 & & \\ & & 1/2 & 1 & 1/2 \\ & & & \cdots & \end{pmatrix},$$

$$R^T = \begin{pmatrix} 1/2 & & \\ 1 & & \\ 1/2 & 1/2 & \\ & 1 & \\ & 1/2 & \\ & & \cdots \end{pmatrix}.$$

The error propagation matrix for the coarse grid correction is given by

$$E_c = I - R^T A_H^{-1} R A_h.$$

We will not directly diagonalize E_c; instead we first convert it to (2×2) block diagonal form and then calculate the eigenvalues of the 2×2 blocks.

Let Q_h denote the eigenvectors of A_h in a particular "nonstandard" order

$$Q_h = [e^1 \; e^n \; e^2 \; e^{n-1} \; \cdots \; e^N \; e^{N+2} \; e^{N+1}].$$

We need a similar matrix for the coarse grid operator A_H.

$$Q_H = [e_H^1 \; e_H^2 \; \cdots \; e_H^N].$$

Note that in this case the eigenvectors are kept in their "natural" ordering. Also

$$Q_H^{-1} A_H Q_H = \begin{pmatrix} \frac{4}{H}\sin^2(\frac{\pi H}{2}) & & \\ & \cdots & \\ & & \frac{4}{H}\sin^2(\frac{N\pi H}{2}) \end{pmatrix}.$$

The reason for choosing the nonstandard ordering of the eigenvectors in Q_h is that with this particular ordering

$$Q_H^{-1} R Q_h = \begin{pmatrix} \cos(\frac{\pi h}{2}) & \cos(\frac{\pi h}{2}) & & & \\ & & \cos(\frac{2\pi h}{2}) & \cos(\frac{2\pi h}{2}) & \\ & & & & \cdots \end{pmatrix}.$$

Thus

$$Q_h^{-1} E_c Q_h = I - (Q_h^{-1} R^T Q_H)(Q_H^{-1} A_H^{-1} Q_H)(Q_H^{-1} R Q_h)(Q_h^{-1} A_h Q_h),$$

$$= \begin{pmatrix} \sin^2(\frac{\pi h}{2}) & \cos^2(\frac{\pi h}{2}) & & & \\ \sin^2(\frac{\pi h}{2}) & \cos^2(\frac{\pi h}{2}) & & & \\ & & \sin^2(\frac{2\pi h}{2}) & \cos^2(\frac{2\pi h}{2}) & \\ & & \sin^2(\frac{2\pi h}{2}) & \cos^2(\frac{2\pi h}{2}) & \\ & & & & \cdots \end{pmatrix}.$$

In general, the block in the lower right corner of $Q_h^{-1} E_c Q_h$ may be a 1×1 block, rather than 2×2. Since it plays no important role in the analysis we will ignore it.

The eigenvectors of the blocks are given by

$$\begin{pmatrix} 1 \\ 1 \end{pmatrix} \quad \text{and} \quad \begin{pmatrix} \cos^2(\frac{i\pi h}{2}) \\ -\sin^2(\frac{i\pi h}{2}) \end{pmatrix},$$

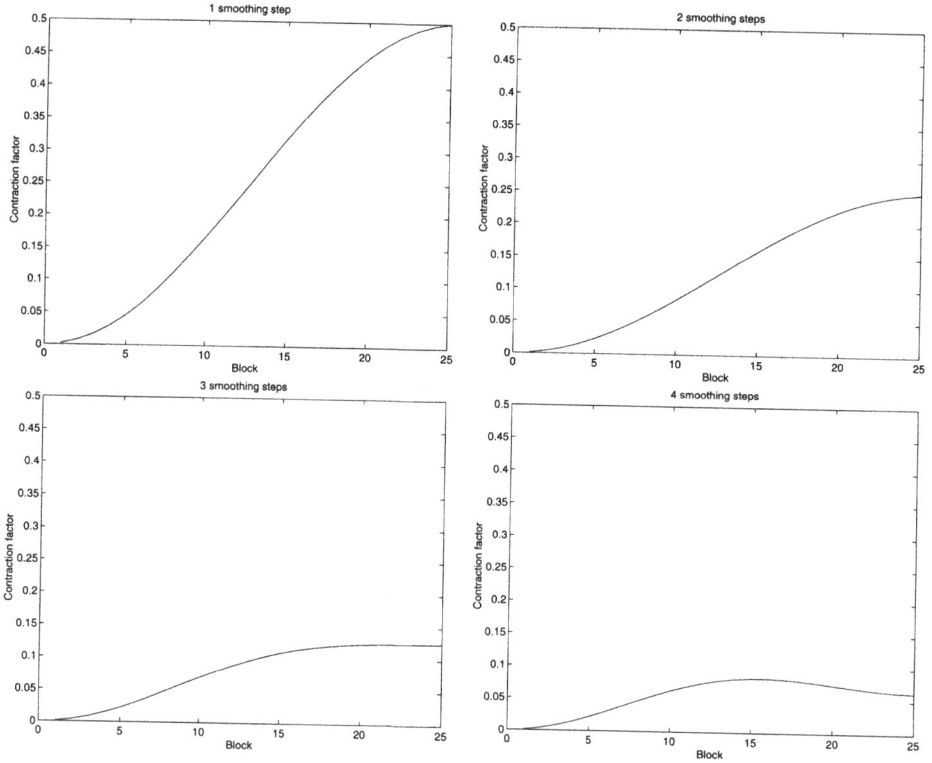

Figure 2.6. Contraction Rate as Function of Subblock

with corresponding eigenvalues of 1 and 0. Thus we can conclude that the eigenvectors of E_c with eigenvalue of 1 are given by

$$e^i + e^{n-i+1},$$

while the eigenvectors of E_c with eigenvalue of 0 are given by

$$f^i = \cos^2\left(\frac{i\pi h}{2}\right) e^i - \sin^2\left(\frac{i\pi h}{2}\right) e^{n-i+1}.$$

For $i < n/4$ the term $\cos^2(i\pi h/2)$ is near 1, while $\sin^2(i\pi h/2)$ is small. Thus the annihilated vectors f^i approximate the low frequency modes of the error. So the coarse grid correction does a good job on the low frequency (smooth) error.

We thus have a quantitative measure of the relative effect of Jacobi smoothing and the coarse grid correction on different frequency modes of the error.

To understand the interaction between the coarse grid correction and the Jacobi smoothing in the multiplicative two level method, we write the error propagation matrix for ν smoothing steps followed by the coarse grid correction as

$$E = E_c E_s^{\nu}.$$

Using the diagonalization given above, E may be converted to

$$Q_h^{-1} E Q_h = \begin{pmatrix} \sin^2(\frac{\pi h}{2}) & \cos^2(\frac{\pi h}{2}) & & & \\ \sin^2(\frac{\pi h}{2}) & \cos^2(\frac{\pi h}{2}) & & & \\ & & \sin^2(\frac{2\pi h}{2}) & \cos^2(\frac{2\pi h}{2}) & \\ & & \sin^2(\frac{2\pi h}{2}) & \cos^2(\frac{2\pi h}{2}) & \\ & & & & \cdots \end{pmatrix}^{\nu}$$

$$\cdot \begin{pmatrix} \cos^2(\frac{\pi h}{2}) & & & & \\ & \sin^2(\frac{\pi h}{2}) & & & \\ & & \cos^2(\frac{2\pi h}{2}) & & \\ & & & \sin^2(\frac{2\pi h}{2}) & \\ & & & & \cdots \end{pmatrix}.$$

In Figure 2.6 we plot the eigenvalues of the subblocks as a function of the frequency i for several smoothing steps, $\nu = 1, 2, 3, 4$. The condition numbers for the multiplicative method used as a preconditioner are given by 2, $\frac{4}{3}$, 1.1429, and 1.0090.

In the analysis of the condition number for the additive method we write the preconditioned problem as

$$BA = \left(R^T A_H^{-1} R + \frac{h}{2} I \right) A_h = 2I - E_c - E_s.$$

Thus

$$Q_h^{-1} B A_h Q_h = \begin{pmatrix} 1 & -\cos^2(\frac{\pi h}{2}) & & & \\ -\sin^2(\frac{\pi h}{2}) & 1 & & & \\ & & 1 & -\cos^2(\frac{2\pi h}{2}) & \\ & & -\sin^2(\frac{2\pi h}{2}) & 1 & \\ & & & & \cdots \end{pmatrix}.$$

The eigenvalues of these blocks are bounded between $\frac{1}{2}$ and $\frac{3}{2}$; thus the condition number for the additive scheme is bounded by 3.

This type of analysis may be carried over into two or three dimensions, except that the subblocks are now 4×4 or 8×8. It is also sometimes possible to calculate the eigenvalues of the subblocks explicitly.

Notes and References for Section 2.7

- The material in this section is based on Section 10.3 in Hackbusch's book on iterative methods [Hac94]. Fourier analysis of problems of this type has been known since the 1970s.

2.8 Variational Formulation

The mathematical convergence analysis for two level methods is somewhat simplified when the coarse grid correction is a projection of the error onto a subspace of the solution space. Even when the correction is only an "approximate" projection, insight into the convergence of the method may still be obtained.

2.8.1 Nested Spaces

In this section we will demonstrate that for conforming C^0, Lagrangian, finite elements the coarse grid corrections are, in fact, projections of the error onto a "coarse" subspace of the solution space. This result will be used in Chapter 5 to determine the convergence rate for a class of overlapping Schwarz two level methods.

Define a triangulation of the domain Ω with elements of diameter $O(H)$; then refine that triangulation (perhaps several times) to obtain a new triangulation with elements of diameter $O(h)$; see Figure 2.7. Consider the space of continuous, piecewise Lagrangian polynomials of degree p on the fine elements, V^h. On the coarse elements construct the space of continuous, piecewise polynomials of degree $q \leq p$, V^H. Then V^H is the **coarse subspace of the fine space** V^h, that is, $V^H \subset V^h$.

Let $\{\phi_i\}$ denote the usual nodal basis for V^h and $\{\psi_k\}$ denote the usual nodal basis of V^H. Thus any function in V^h can be represented as

$$\boldsymbol{u} = \sum_i u_i \phi_i$$

and similarly any function in V^H can be written as

$$\boldsymbol{u}_C = \sum_k u_{C_k} \psi_k.$$

Since $V^H \subset V^h$, we can express the coarse grid basis functions, $\{\psi_k\}$, as linear combinations of the fine grid basis functions $\{\phi_j\}$,

$$\psi_k = \sum_j R_{kj} \phi_j.$$

Then

$$\begin{aligned}
\boldsymbol{u}_C &= \sum_k u_{C_k} \psi_k, \\
&= \sum_k u_{C_k} \sum_j R_{kj} \phi_j, \\
&= \sum_j (\sum_k u_{C_k} R_{kj}) \phi_j, \\
&= \sum_j [R^T u_C]_j \phi_j.
\end{aligned}$$

So we see that R^T is exactly the matrix operator which performs interpolation from the coefficients of the coarse grid function to the coefficients on the fine grid.

Define the projection, $\boldsymbol{P}_0\boldsymbol{u}$, of a function $\boldsymbol{u} \in V^h$ onto the subspace V^H by

$$a(\boldsymbol{P}_0\boldsymbol{u}, \boldsymbol{v}) = a(\boldsymbol{u}, \boldsymbol{v}) \qquad \boldsymbol{P}_0\boldsymbol{u} \in V^H, \ \forall \boldsymbol{v} \in V^H.$$

We shall now derive explicit matrix representation of the operator $\boldsymbol{P}_0$. Let $\boldsymbol{w}_C = \sum_k w_{C_k} \psi_k = \boldsymbol{P}_0\boldsymbol{u}$; then

$$a(\boldsymbol{w}_C, \psi_j) = a(\boldsymbol{u}, \psi_j) \quad \forall \psi_j,$$

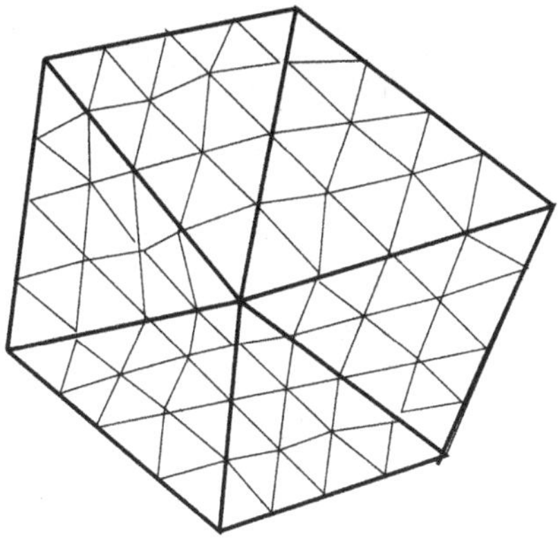

Figure 2.7. Triangulation into Subdomains and Elements

$$
\begin{aligned}
\sum_k w_{C_k} a(\psi_k, \psi_j) &= \sum_k u_k a(\phi_k, \psi_j) \quad \forall \psi_j,\\
&= \sum_k u_k \sum_l R_{jl} a(\phi_k, \phi_l),\\
&= \sum_k u_k \sum_l R_{jl} A_{lk},\\
A_C w_C &= RAu,\\
w_C &= A_C^{-1} RAu.
\end{aligned}
$$

The values of w_C must be interpolated onto the fine grid so we obtain

$$P_0 = R^T A_C^{-1} RA.$$

Note that

$$
\begin{aligned}
(A_C)_{kj} &= a(\psi_j, \psi_k),\\
&= \sum_l \sum_i R_{lk} R_{ij} a(\phi_i, \phi_l),\\
&= (RAR^T)_{kj}.
\end{aligned}
$$

Thus we have shown that for conforming C^0 finite elements, the coarse grid correction is a projection of the error onto a "coarse subspace."

2.8.2 Nonnested Subspaces

In the case of nonnested subspaces the space V^H is not a subspace of V^h. This means that the basis functions of V^H, ψ_k, cannot be written as a linear combination of the basis functions ϕ_j. Let $\boldsymbol{I}$ be a linear operator from V^H to V^h; we may think of $\boldsymbol{I}$ as an interpolationlike operator. Now $\boldsymbol{I}\psi_k$ is in V^h and hence can be expressed as a linear combination of basis functions of ϕ_j. Define R by

$$\boldsymbol{I}\psi_k = \sum_j R_{kj}\phi_j.$$

Now if

$$\boldsymbol{u}_C = \sum_k u_{C_k} \psi_k$$

then

$$\begin{aligned} \boldsymbol{u} &= \boldsymbol{I}\boldsymbol{u}_C, \\ &= \sum_k u_{C_k} \boldsymbol{I}\psi_k, \\ &= \sum_j \phi_j \sum_k R_{kj} u_{C_k}, \\ &= \sum_j [R^T u_C]_j \phi_j. \end{aligned}$$

Therefore $u_j = [R^T u_C]_j$, that is, $u = R^T u_C$. Hence the matrix R^T still "interpolates" the coefficients from the "coarse" space to the "fine" space. Note that it now depends on the linear operator $\boldsymbol{I}$.

There are two natural ways to define the coarse grid correction operator for nonnested spaces.

Method 1: Define $\tilde{\boldsymbol{T}}_0\boldsymbol{u}$, $\boldsymbol{u} \in V^h$ by

$$a(\tilde{\boldsymbol{T}}_0\boldsymbol{u}, \boldsymbol{v}) = a(\boldsymbol{u}, \boldsymbol{I}\boldsymbol{v}) \qquad \tilde{\boldsymbol{T}}_0\boldsymbol{u} \in V^H, \ \forall \boldsymbol{v} \in V^H.$$

We will refer to $\tilde{\boldsymbol{T}}_0$ as a projectionlike operator. We now derive an explicit formula for the matrix representation of $\boldsymbol{I}\tilde{\boldsymbol{T}}_0$ in the $\{\phi_j\}$ basis. (The $\boldsymbol{I}$ is needed to interpolate the function back into V^h from V^H.) Let

$$\boldsymbol{w}_C = \tilde{\boldsymbol{T}}_0\boldsymbol{u} = \sum_k w_{C_k} \psi_k$$

and

$$\boldsymbol{u}_C = \boldsymbol{I}\boldsymbol{w}_C = \boldsymbol{I}\tilde{\boldsymbol{T}}_0\boldsymbol{u} = \sum_j u_{C_j} \phi_j.$$

Then

$$\begin{aligned} a(\boldsymbol{w}_C, \psi_j) &= a(\boldsymbol{u}, \boldsymbol{I}\psi_j) && \forall \psi_j, \\ \textstyle\sum_k w_{C_k} a(\psi_k, \psi_j) &= \textstyle\sum_k u_k a(\phi_k, \boldsymbol{I}\psi_j) && \forall \psi_j, \\ &= \textstyle\sum_k u_k \sum_l R_{jl} a(\phi_k, \phi_l) && \forall j, \\ &= \textstyle\sum_l R_{jl} \sum_k A_{lk} u_k && \forall j, \\ &= (RAu)_j. \end{aligned}$$

Solving for w_C gives

$$w_C = \tilde{A}_C^{-1} RAu,$$

with

$$(\tilde{A}_C)_{ij} = a(\psi_j, \psi_i).$$

In general, for Method 1, $\tilde{A}_C \neq RAR^T$. When we interpolate back to the fine grid we obtain

$$u_C = R^T \tilde{A}_C^{-1} RAu.$$

Finally we consider the case when the bilinear form $a_0(\cdot, \cdot)$ on the "coarse" space is not equal to $a(\cdot, \cdot)$. Then by the same type of calculation as above

$$u_C = R^T \tilde{A}_C^{-1} RAu$$

where

$$(\tilde{A}_C)_{ij} = a_0(\psi_j, \psi_i).$$

Method 2: In the second approach, we introduce an auxiliary **nested** coarse grid space, $\boldsymbol{I}V^H \subset V^h$ and then define $\boldsymbol{T}_0$ as was done in the case of nested spaces, for $\boldsymbol{u} \in V^h$,

$$a(\boldsymbol{T}_0\boldsymbol{u}, \boldsymbol{v}) = a(\boldsymbol{u}, \boldsymbol{v}) \qquad \boldsymbol{T}_0\boldsymbol{u} \in \boldsymbol{I}V^H, \ \forall v \in \boldsymbol{I}V^H.$$

Since $\boldsymbol{T}_0\boldsymbol{u} \in \boldsymbol{I}V^H$ it may be represented as

$$\begin{aligned} \boldsymbol{u}_C &= \boldsymbol{T}_0\boldsymbol{u}, \\ &= \sum_k w_{C_k} \boldsymbol{I}\psi_k, \\ &= \sum_j [Rw_C]_j \phi_j, \end{aligned}$$

or, in other words, $u_C = R^T w_C$.

To derive the matrix representation of $\boldsymbol{T}_0$ we observe

$$\begin{aligned} a(\sum_k w_{C_k} \boldsymbol{I}\psi_k, \boldsymbol{I}\psi_j) = a(\boldsymbol{u}, \boldsymbol{I}\psi_j) \quad \forall j, \\ &= a(\sum_i u_i\phi_i, \sum_l R_{jl}\phi_l), \\ &= \sum_i \sum_l R_{jl} a(\phi_i, \phi_l) u_i, \\ &= \sum_i \sum_l R_{jl} A_{li} u_i, \\ &= (RAu)_j. \end{aligned}$$

Hence,

$$\bar{A}_C w_C = RAu$$

or

$$u_C = R^T \bar{A}_C^{-1} R A u,$$

where

$$(\bar{A}_C)_{ij} = a(\boldsymbol{I}\psi_j, \boldsymbol{I}\psi_i),$$
$$= (RAR^T)_{ij}.$$

In the case when the bilinear form used on the coarse grid does not equal $a(\cdot, \cdot)$, then

$$(\bar{A}_C)_{ij} = a_0(\boldsymbol{I}\psi_j, \boldsymbol{I}\psi_i).$$

Note that the only difference between Methods 1 and 2 is that the coarse grid stiffness matrix is different, whereas the algebraic restriction and interpolation remain exactly the same. Also note that Method 2 is identical to Method 1 with the "approximate" bilinear form $a_0(\cdot, \cdot) = a(\boldsymbol{I}\cdot, \boldsymbol{I}\cdot)$.

Notes and References for Section 2.8

- For the case of nonnested two level methods the presentation given for Method 1 is similar to the approach taken in Bramble [Bra93].
- The case where the nonnestedness of the subspaces is due to nonnested grids has been studied by X. Cai [Cai95] and Chan, Smith, and Zou [Cha94d].

3
Multilevel Algorithms

FOR PROBLEMS ON FINE GRIDS with large numbers of unknowns there are two extreme choices of subdomain sizes. One may have a large number of reasonably small subdomains or a small number of large subdomains. In the former case the global, coarse grid problem which must be solved is large; in the latter, the subdomain problems are large, often too large to be solved on a single processor. Therefore for large problems two level algorithms may be inadequate, simply as a result of memory considerations. Another, slightly more subtle reason to consider multilevel methods is that they may dramatically reduce the total amount of computational work needed to solve the problem to a particular accuracy. This is discussed in Section 3.6.

It is natural to solve the resulting large problems recursively by another application of the two level preconditioner. The resulting algorithm is a **multilevel Schwarz method.** As with the two level methods, there is a great deal of freedom in choosing the order in which the subdomain solves are applied as well as subdomain sizes, and so on. Certain choices will result in many of the standard multigrid and multilevel algorithms.

In most multigrid literature, these methods are not usually presented as preconditioners for Krylov subspace methods, rather as complete iterative methods in themselves. Historically, multigrid methods were developed before most Krylov subspace methods and before preconditioning for Krylov subspace methods was widely understood. Another reason for not using Krylov subspace accelerators is that for certain problems, multigrid algorithms are so powerful that the extra improvement Krylov subspace methods produce is insignificant. All algorithms in this book are presented as preconditioners; thus several of the methods which are, in fact, standard multigrid methods may appear in a slightly different form from what many multigrid practitioners are used to. This chapter is not intended to replace texts on multigrid methods, but to complement them and demonstrate the intertwining relationship between multigrid and domain decomposition algorithms.

3.1 Additive Multilevel Schwarz Methods

This section introduces the multilevel additive Schwarz methods. These are natural extensions of the two level additive Schwarz methods. Essentially the coarse grid solve introduced in the previous chapter is itself replaced by a two level preconditioner. In order to explain this more completely we need to introduce some additional notation.

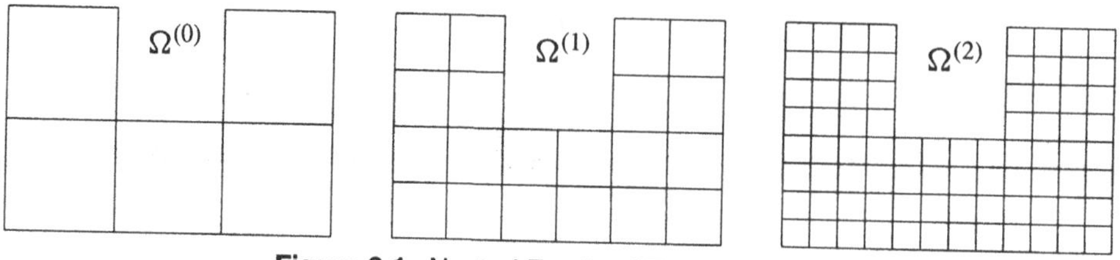

Figure 3.1. Nested Family of Triangulations

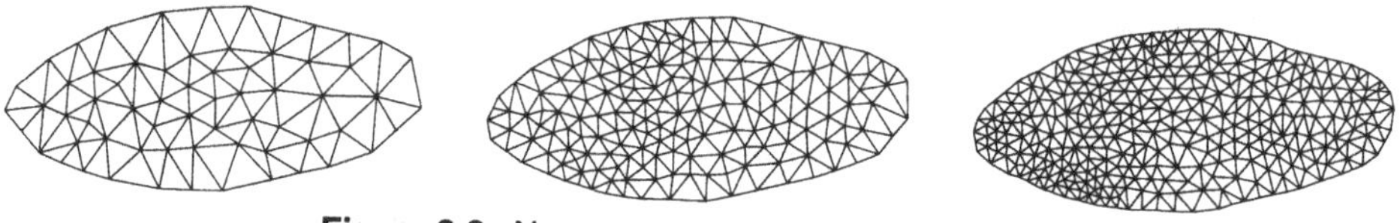

Figure 3.2. Nonnested Family of Triangulations

Assume we have a domain Ω and a (possibly nonnested) family of triangulations $\Omega^{(i)}$, for $i = 0, \ldots, j$, with characteristic diameters $h^{(i)}$; see Figures 3.1 and 3.2 for two sample domains with nested and nonnested triangulations. We also assume that we have discretized the PDE on each level of the triangulation. Let $A^{(i)}$ be the matrix obtained by discretizing the PDE on the ith level of triangulation. In addition, let the vector $u^{(i)}$ contain the coefficients of $\boldsymbol{u}$ associated with the triangulation $\Omega^{(i)}$. Throughout this chapter the superscript (i) will denote a quantity associated with the ith level. $\Omega^{(0)}$ represents the coarsest level of discretization and $\Omega^{(j)}$ the finest grid. We decompose the grid $\Omega^{(i)}$ into $N^{(i)}$ subdomains.

Let $R_k^{(i)}$, for $k = 1, \ldots, N^{(i)}$, be the restriction operators for the domains $\Omega_k^{(i)}$; that is, $R_k^{(i)} u^{(i)}$ returns the vector of all the coefficients associated with domain $\Omega_k^{(i)}$. In addition, define $R^{(i)}$ to be the restriction to the ith level, that is, from the next finer level $i+1$ to level i. In practice, the matrix $R^{(i)}$ is often defined from the requirement that $R^{(i)^T}$ is a reasonable interpolation from grid i to grid $i+1$. Similarly, the coarser level stiffness matrices are often Galerkin, that is, derived from the fine grid matrix by using the relation $A^{(i)} = R^{(i)} A^{(i+1)} R^{(i)^T}$.

The matrix $A_k^{(i)} = R_k^{(i)} A^{(i)} R_k^{(i)^T}$ is the subblock of $A^{(i)}$ associated with domain $\Omega_k^{(i)}$. In most applications, the matrices $A_k^{(i)}$ will be rather small. For instance, its dimension may be around 25. In the extreme case the dimensions of the subspaces can be chosen to be 1.

We keep in mind that the subdomains $\Omega_k^{(i)}$, just as in the previous chapters, overlap each other. This overlap is often small; in the extreme case when $A_k^{(i)}$ represents a single node there is still an overlap. To see this, consider two neighboring nodes connected by an edge in Figure 3.3. The two corresponding subdomains, Ω_1 and Ω_2, consist of the support of the basis function associated with the two center nodes; thus the overlap consists of the two intermediate triangles.

The standard two level additive Schwarz preconditioner, with a large coarse problem,

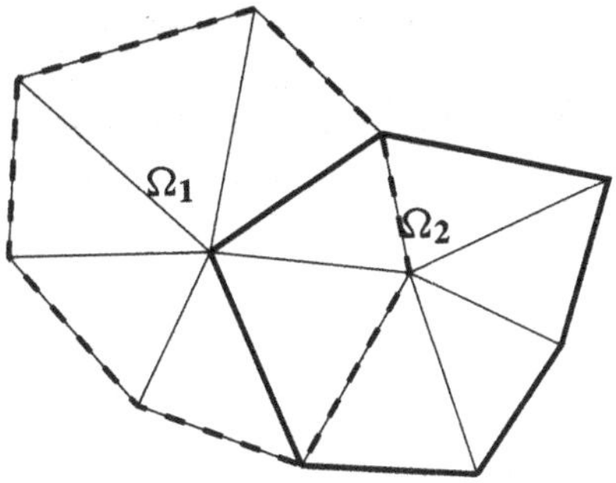

Figure 3.3. Minimal Overlap

is given by

$$B^{(j)} = \sum_{k=1}^{N^{(j)}} R_k^{(j)^T} A_k^{(j)^{-1}} R_k^{(j)} + R^{(j-1)^T} A^{(j-1)^{-1}} R^{(j-1)}. \tag{3.1}$$

The subproblems involving $A_k^{(j)}$ are small and can be solved quickly with either a direct or an iterative solver. The problem involving $A^{(j-1)}$ may, however, be quite large and so we may not wish to solve it exactly. Fortunately, we have a good preconditioner for $A^{(j-1)}$, that is, something that approximates $A^{(j-1)^{-1}}$ well. We can simply use the two level additive overlapping Schwarz preconditioner to replace $A^{(j-1)^{-1}}$,

$$A^{(j-1)^{-1}} \approx B^{(j-1)} = \sum_{k=1}^{N^{(j-1)}} R_k^{(j-1)^T} A_k^{(j-1)^{-1}} R_k^{(j-1)} + R^{(j-2)^T} A^{(j-2)^{-1}} R^{(j-2)}.$$

So the three level additive preconditioner $B^{(j)}$ may be written as

$$\sum_{k=1}^{N^{(j)}} R_k^{(j)^T} A_k^{(j)^{-1}} R_k^{(j)}$$
$$+ R^{(j-1)^T} \left(\sum_{k=1}^{N^{(j-1)}} R_k^{(j-1)^T} A_k^{(j-1)^{-1}} R_k^{(j-1)} + R^{(j-2)^T} A^{(j-2)^{-1}} R^{(j-2)} \right) R^{(j-1)}.$$

We can apply this technique recursively to approximate $A^{(j-2)^{-1}}$, and so on, until reaching $A^{(0)^{-1}}$.

The application of the additive multilevel Schwarz preconditioner to a vector $u = Br$ can be written as in Algorithm 3.1.1.

If there is only a single domain on the coarsest level, that is, $N^{(0)} = 1$, this is equivalent to using an exact solver on that grid. Since the coarsest grid may represent only a small number of unknowns, this is commonly done.

Note the recursive nature of the application of the restriction and interpolation. To express this as a one step preconditioner, define $\bar{R}^{(i)} = \prod_{l=i}^{j-1} R^{(l)}$ and $\bar{R}_k^{(i)} = R_k^{(i)} \bar{R}^{(i)}$ consistently with our previous notation. Figure 3.4 shows the relationship of the restriction operators introduced in this chapter. That is, $\bar{R}_k^{(i)}$ maps from the finest

Algorithm 3.1.1: Additive Multilevel Schwarz Preconditioner

$$
\begin{aligned}
r^{(j)} &\leftarrow r \\
r^{(j-1)} &\leftarrow R^{(j-1)} r^{(j)} \\
&\dots \\
r^{(0)} &\leftarrow R^{(0)} r^{(1)} \\
u^{(i)} &\leftarrow \sum_{k=1}^{N^{(i)}} R_k^{(i)^T} A_k^{(i)^{-1}} R_k^{(i)} r^{(i)} \quad \forall i = 0, \dots, j \\
u^{(1)} &\leftarrow u^{(1)} + R^{(0)^T} u^{(0)} \\
&\dots \\
u^{(j)} &\leftarrow u^{(j)} + R^{(j-1)^T} u^{(j-1)} \\
u &\leftarrow u^{(j)}.
\end{aligned}
$$

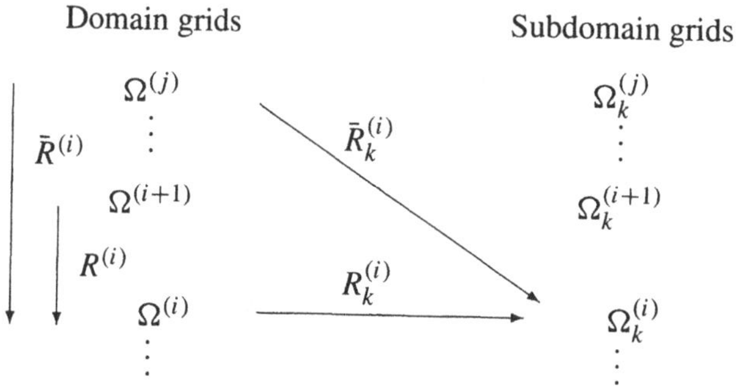

Figure 3.4. The Restriction Operators. Note that the Superscript Always Denotes the Target Grid Level and that a Bar Shows that We Restrict from the Finest Grid. Conversely, the Transpose of These Operators Interpolates from the Grid Level Indicated by the Superscript.

level to the kth subdomain on the ith level. Then the preconditioner may be written as

$$B = \sum_{i=0}^{j} \sum_{k=1}^{N^{(i)}} \bar{R}_k^{(i)^T} A_k^{(i)^{-1}} \bar{R}_k^{(i)}. \tag{3.2}$$

Using the language introduced in Section 2.2, we are applying the additive Schwarz smoother on all levels simultaneously.

In Algorithm 3.1.1 presented above, all of the local solves on all of the levels may be applied in parallel. However, the restrictions and interpolations are calculated sequentially, while in the purely parallel form (3.2), even the interpolation and restriction may be calculated in parallel. The floating point costs of the two approaches are quite different. This difference is easily estimated. For simplicity, we consider only the case

Algorithm 3.1.2: Multilevel Diagonal Scaling Preconditioner

$$B = \bar{R}^{(0)^T} A^{(0)^{-1}} \bar{R}^{(0)} + \sum_{i=1}^{j-1} \bar{R}^{(i)^T} D^{(i)^{-1}} \bar{R}^{(i)} + D^{(j)^{-1}}.$$

$h^{(i)}/h^{(i-1)} = \frac{1}{2}$. On a one dimensional grid the number of grid points is $2^i + 1$ on grid level i if the coarsest grid $\Omega^{(0)}$ has two points. Similarly, the number of grid points on a d-dimensional grid $\Omega^{(i)}$ is $O(2^{id})$. The cost of the interpolation (restriction) between two adjacent grids depends on the number of new points introduced on the finer grid. Thus, the cost of the recursive (sequential) approach can be estimated as

$$\sum_{i=1}^{j}(2^{id} - 2^{(i-1)d}) = 2^{jd} - 1 = O(N),$$

where N is the number of unknown coefficients on the finest grid. In the completely parallel form, each interpolation (restriction) is computed between levels i and j. The number of levels j is $O(\log(N))$, and the total cost can therefore be estimated as

$$\sum_{i=0}^{j-1} 2^{jd} = O(j2^{jd}) = O(N\log(N)).$$

It is important to understand that whether the interpolation and restriction are calculated recursively or not, the preconditioner is still a purely additive preconditioner. The multiplicative forms of the preconditioner are discussed later in this chapter.

There are several special cases of additive multilevel Schwarz methods that deserve our attention. We introduce these in the next three sections.

3.1.1 Multilevel Diagonal Scaling

An important special case of additive multilevel Schwarz is obtained when we use minimal size subdomains. That is, let $R_k^{(i)}$ return the coefficients (single coefficient for a scalar PDE) related to a single node; then

$$D^{(i)} = \sum_{k=1}^{N^{(i)}} R_k^{(i)^T} A_k^{(i)} R_k^{(i)} \tag{3.3}$$

corresponds to the (block, for multicomponent problems) diagonal of the matrix $A^{(i)}$. Let us pull out the first and last entry in the sum (3.2), noting that the restriction operators for the extreme grid levels have special properties. By factoring $\bar{R}_k^{(i)} = R_k^{(i)} \bar{R}^{(i)}$ and using (3.3) we obtain a new preconditioner. It is called the **multilevel diagonal scaling** preconditioner and we can write it as in Algorithm 3.1.2. We have assumed that $N^{(0)} = 1$, that is, the coarsest grid consists of only a single subdomain that we are prepared to solve directly. Observe that with this preconditioner one only needs to know the diagonal elements of the matrix $A^{(i)}$ except on the coarsest level, where the entire stiffness matrix $A^{(0)}$ must be known. It is possible to use a simple diagonal solve on the coarsest grid as well, but this may lead to an inferior rate of convergence.

Computational Results 3.1.1: Convergence Behavior for Multilevel Diagonal Scaling

Purpose: Demonstrate the convergence behavior of multilevel diagonal scaling.

PDE: The Poisson equation,

$$\begin{aligned} -\Delta \boldsymbol{u} &= xe^y \quad \text{in } \Omega, \\ \boldsymbol{u} &= -xe^y \quad \text{on } \partial\Omega, \end{aligned}$$

with Dirichlet boundary conditions.

Domain: Unit square.

Discretization: Piecewise linear finite elements.

Calculations: Two different sets of results are presented here. The first table shows the convergence behavior as the size of the finest grid is held constant, adjusting the number of levels (and consequently the size of the coarse grid) for two sizes. In all cases, the Krylov subspace accelerator method used was GMRES, with a restart of 10. The coarse grid problem was solved exactly, using a direct sparse solver.

Iteration Counts for 10^{-2}
Reduction in Error

N_{fine}	Levels	Iterations
33	2	7
33	3	8
33	4	14
65	2	7
65	3	8
65	4	14
65	5	23

The second table shows the convergence behavior as the coarse grid is held constant and the number of levels is increased (thus increasing the size of the fine grid).

Iteration Counts for 10^{-2} Reduction in Error

N_{coarse}	Levels	Iterations
9	2	6
9	3	8
9	4	14
9	5	23
17	2	7
17	3	8
17	4	13

Discussion: These two tables show that the convergence rate is relatively independent of the fine grid size but depends strongly on the number of levels. The second table show how strongly the iteration count depends on the number of levels with a given coarse grid. Though, in fact, the convergence rate is bounded independently of the number of levels (see Section 5.3.4), in practice, for realistic grid sizes (as seen here), the number of iterations does increase with the number of levels.

3.1.2 The BPX Method

For the model operator $L = -\Delta$, and bilinear (trilinear) finite elements on a uniform or quasi-uniform grid in d dimensions, we can estimate the size of the diagonal elements $D^{(i)}$ from (3.3) quite easily. Recall that each entry in the stiffness matrix is built from a fixed number of contributions from $a(\phi_j, \phi_i) = \int_\Omega \nabla\phi_j \nabla\phi_i dx$. When ϕ is linear in each variable, its gradient must have a slope of $O(1/h^{(i)})$ since a basis function varies from 0 to 1 over the element size $h^{(i)}$. The contributions from the integral will therefore have the form $V h^{(i)^{-2}}$ where V is the volume of the element. V is $O(h^{(i)^d})$ for a d-dimensional element; hence we conclude that the diagonal elements must be of the form $O(h^{(i)^{d-2}})$.

We have just shown that $D^{(i)}$ scales like $O(h^{(i)^{d-2}})$. It turns out that this captures the most essential properties for all second order uniformly elliptic problems. Hence, one can use the simplified preconditioner in Algorithm 3.1.3. This preconditioner is referred to as the **BPX** preconditioner or the **multilevel nodal basis preconditioner**. The method has received much attention, which is due to its simplicity and historical importance in relating domain decomposition and multigrid ideas.

The additive multilevel Schwarz preconditioner has very good theoretical convergence properties for scalar, self-adjoint uniformly elliptic PDEs. It can be shown that the number of iterations required to achieve a fixed tolerance is bounded independently of the number of unknowns, see Section 5.3.4. Moreover, if one uses the recursive formulation of the interpolation and restriction, the amount of work required per iteration is proportional to the number of unknowns. The two special cases of the general algorithm, the multilevel diagonal scaling preconditioner as well as the BPX method,

Algorithm 3.1.3: BPX Preconditioner

$$B = \bar{R}^{(0)^T} A^{(0)^{-1}} \bar{R}^{(0)} + \sum_{i=1}^{j-1} h^{(i)^{2-d}} \bar{R}^{(i)^T} \bar{R}^{(i)} + h^{(j)^{2-d}} I.$$

Algorithm 3.1.4: Hierarchical Basis Preconditioner

$$B_H = \bar{R}^{(0)^T} A^{(0)^{-1}} \bar{R}^{(0)} + \sum_{i=1}^{j-1} \bar{R}_H^{(i)^T} D_H^{(i)^{-1}} \bar{R}_H^{(i)} + \bar{R}_H^{(j)^T} D_H^{(j)^{-1}} \bar{R}_H^{(j)}.$$

also have similar good convergence properties. The BPX algorithm, however, uses less explicit information about the operator, and one would expect it, in general, to be less robust as one changes the coefficients of the PDE.

3.1.3 The Hierarchical Basis Method

Another multilevel preconditioner, the **hierarchical basis** method, can be derived from the multilevel nodal basis preconditioner by simply dropping certain terms. Note that the coefficients associated with nodes on the coarser triangulations are dealt with several times by the preconditioner B, both on the coarsest triangulation on which they appear and on all finer triangulations. In order to reduce the amount of computation per iteration we can simply treat each node only once, on the coarsest triangulation in which it appears; see Figure 3.5. Define $\bar{R}_H^{(i)}$ as the submatrix of $\bar{R}^{(i)}$ containing only those rows associated with nodes which have not yet appeared on a coarser triangulation. The hierarchical basis preconditioner is given by

$$B_H = \bar{R}^{(0)^T} A^{(0)^{-1}} \bar{R}^{(0)} + \sum_{i=1}^{j-1} h^{(i)^{2-d}} \bar{R}_H^{(i)^T} \bar{R}_H^{(i)} + h^{(j)^{2-d}} \bar{R}_H^{(j)^T} \bar{R}_H^{(j)}.$$

The matrix $\bar{R}_H^{(j)}$ has the following obvious meaning: it simply returns all the coefficients that have been neglected on all of the previous coarser levels. It is also possible (as with multilevel diagonal scaling) to use the diagonals on each level for scaling. Define $D_H^{(i)}$ to be the submatrix of $D^{(i)}$ containing the entries associated with the nodes selected by $\bar{R}_H^{(i)}$. An alternative hierarchical basis preconditioner may then be written as in Algorithm 3.1.4.

The hierarchical preconditioner does not perform quite as well as the BPX preconditioner in two dimensions. It can be shown that for scalar, self-adjoint elliptic problems in two dimensions the number of iterations required to achieve a fixed tolerance grows as

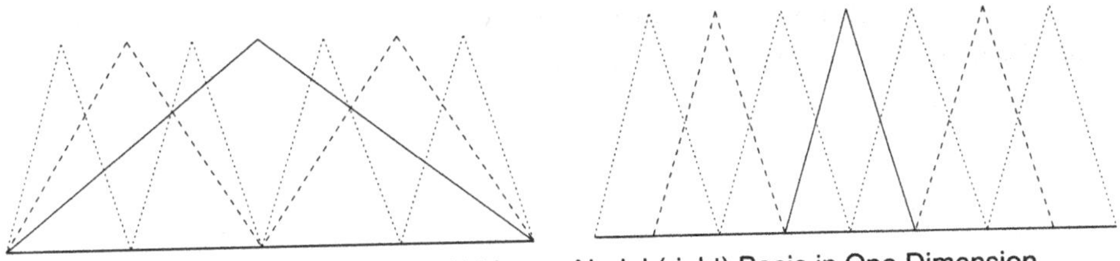

Figure 3.5. A Hierarchical (left) Versus Nodal (right) Basis in One Dimension

$O(1+\log(h^{(0)}/h^{(j)}))$. In three dimensions the convergence is quite poor; the number of iterations required to reach a fixed tolerance grows like $\sqrt{h^{(0)}/h^{(j)}}(1+\log(h^{(0)}/h^{(j)}))$. As with the additive multilevel Schwarz preconditioner, the hierarchical basis preconditioner's interpolation (restriction) may be applied recursively with $O(N)$ work or completely in parallel with $O(N\log(N))$ work.

This preconditioner is called the hierarchical basis method, because it can be viewed as a change to a hierarchical basis followed by a (block) diagonal preconditioning. Define the matrix H by letting the rows of H be first the rows of $\bar{R}_H^{(0)}$, followed by the rows of $\bar{R}_H^{(1)}$, and so on, up to the rows of $\bar{R}_H^{(j)}$. That is,

$$H = \begin{pmatrix} \bar{R}_H^{(0)} \\ \bar{R}_H^{(1)} \\ \cdots \\ \bar{R}_H^{(j)} \end{pmatrix}.$$

The matrix H is the basis change from the usual nodal basis to the hierarchical basis; see Figure 3.5. If D_H is given by

$$D_H = \begin{pmatrix} A^{(0)} & & & \\ & D_H^{(1)} & & \\ & & \cdots & \\ & & & D_H^{(j)} \end{pmatrix},$$

then

$$B_H = H^T D_H^{-1} H.$$

The hierarchical basis preconditioner is simply a change to the hierarchical basis, a (block) diagonal scaling, and then a change back to the nodal basis.

It is also possible to write the multilevel diagonal scaling preconditioner (or the BPX or the additive multilevel Schwarz preconditioner) by using the same compact notation. Let the matrix M be given by

$$M = \begin{pmatrix} \bar{R}^{(0)} \\ \bar{R}^{(1)} \\ \cdots \\ \bar{R}^{(j-1)} \\ I \end{pmatrix}.$$

Note that M has many more rows than columns, while the matrix H is square. Let D_M be the block diagonal matrix

$$D_M = \begin{pmatrix} A^{(0)} & & & \\ & D^{(1)} & & \\ & & \cdots & \\ & & & D^{(j)} \end{pmatrix};$$

then the multilevel diagonal scaling preconditioner is given by

$$B = M^T D_M^{-1} M. \tag{3.4}$$

Thus all of the additive multilevel methods may be viewed in terms of a change (and expansion) of the basis; followed by a block diagonal preconditioning (scaling) and a change back to the original basis; see also Section 3.5.

Notes and References for Section 3.1

- The additive multilevel Schwarz method is due to Dryja and Widlund [Dry91].
- The algorithm and its convergence theory were developed more completely by Zhang [Zha91]. Zhang also introduced the term **multilevel diagonal scaling.**
- The BPX algorithm, due to Bramble, Pasciak, and Xu [Bra90], is essentially a special case of the additive multilevel Schwarz method, though it was derived in a different manner.
- Oswald [Osw92b] first proved the optimality of the BPX algorithm by using Besov spaces. Zhang [Zha91] had earlier proved a slightly weaker result, though stronger than the result of Bramble, Pasciak, and Xu.
- The idea of using a hierarchical basis instead of the standard nodal basis in order to obtain a better preconditioned system is very intuitive. It was Yserentant [Yse86a], [Yse86b], [Yse86c] who first developed proofs for the convergence rates of hierarchical basis methods in two dimensions. See also [Yse90], where Yserentant compares the hierarchical basis method with the BPX algorithm.
- See Ong [Ong89] for a discussion of the convergence of the hierarchical basis method in three dimensions.

3.2 Multiplicative Multilevel Schwarz Methods

As discussed in Chapters 1 and 2, the difference between multiplicative and additive preconditioners is merely that in the multiplicative form the residual is updated between substeps of the preconditioner. With multilevel Schwarz methods, there are many ways to choose the substeps and the order in which they are traversed. The basic building blocks for the additive and multiplicative methods are the same; the only difference is the order of their application.

3.2.1 Multiplicative between Levels, Additive in Levels

We begin with a straightforward method where the updates on a particular level are in parallel but we move sequentially through the levels. Let $u^{(i)}$ denote the approximate solution at the ith substep of the preconditioner.

$$\begin{aligned}
u^{(1)} &\leftarrow \sum_{k=1}^{N^{(j)}} R_k^{(j)^T} A_k^{(j)^{-1}} R_k^{(j)} r, \\
u^{(2)} &\leftarrow u^{(1)} + \sum_{k=1}^{N^{(j-1)}} \bar{R}_k^{(j-1)^T} A_k^{(j-1)^{-1}} \bar{R}_k^{(j-1)} (r - Au^{(1)}), \\
&\dots \\
u &\leftarrow u^{(j)} + \sum_{k=1}^{N^{(0)}} \bar{R}_k^{(0)^T} A_k^{(0)^{-1}} \bar{R}_k^{(0)} (r - Au^{(j)}).
\end{aligned} \tag{3.5}$$

In most applications the problem on the coarsest grid is solved with a direct solver; in that case the final line of the algorithm may be written as

$$u \leftarrow u^j + \bar{R}^{(0)^T} A^{(0)^{-1}} \bar{R}^{(0)} (r - Au^{(j)}).$$

This is equivalent to having a single subdomain on the coarsest grid. To symmetrize this method, we must traverse back up through the levels.

There are two problems with this algorithm as presented. First, the interpolation (restriction) requires $O(N \log(N))$ work for each application of the preconditioner and, second, the entire residual must be calculated on each level, thus $j \sim \log(N)$ times per application of the preconditioner. Hence, the application of the preconditioner requires a minimum of $O(N \log(N))$ operations, while we desire the optimal $O(N)$.

These problems may be overcome by applying the interpolation (restriction) recursively and updating the residual only on the level where it is needed. Thus the application of the preconditioner to a vector r can be computed as shown in Algorithm 3.2.1, and its symmetrized form may be written as in Algorithm 3.2.2. Many different variations of Algorithms 3.2.1 and 3.2.2 are possible. One may, for instance, apply the additive correction on a level several times before moving to the next level.

When the restriction and interpolation are determined by a Galerkin condition, that is, $A^{(i-1)} = R^{(i-1)} A^{(i)} R^{(i-1)^T}$ (see Section 2.4), the recursive and nonrecursive preconditioners give results that are identical in exact arithmetic. We conclude this section by proving this equivalence. In Chapter 5, we will discuss how this result may be used to obtain convergence proofs for several of the methods by using the same mathematical estimates.

To demonstrate the equivalence of the recursive and nonrecursive forms we need to show that

$$r^{(i)} = \bar{R}^{(i)} (r - Au^{(j-i)}) \quad \forall i = j-1, \dots, 0, \tag{3.7}$$

where $r^{(i)}$ is obtained from (3.6) and $u^{(j-i)}$ is from (3.5). We note that the superscript here refers to the sequence of partial steps contained in a complete iteration step.

Algorithm 3.2.1: Multiplicative Multilevel Schwarz Preconditioner

$$
\begin{aligned}
u^{(j)} &\leftarrow \sum_{k=1}^{N^{(j)}} R_k^{(j)^T} A_k^{(j)^{-1}} R_k^{(j)} r \\
r^{(j-1)} &\leftarrow R^{(j-1)}(r - Au^{(j)}) \\
u^{(j-1)} &\leftarrow \sum_{k=1}^{N^{(j-1)}} R_k^{(j-1)^T} A_k^{(j-1)^{-1}} R_k^{(j-1)} r^{(j-1)} \\
r^{(j-2)} &\leftarrow R^{(j-2)}(r^{(j-1)} - A^{(j-1)}u^{(j-1)}) \\
&\cdots \\
u^{(0)} &\leftarrow \sum_{k=1}^{N^{(0)}} R_k^{(0)^T} A_k^{(0)^{-1}} R_k^{(0)} r^{(0)} \\
u^{(1)} &\leftarrow u^{(1)} + R^{(0)^T} u^{(0)} \\
&\cdots \\
u^{(j)} &\leftarrow u^{(j)} + R^{(j-1)^T} u^{(j-1)} \\
u &\leftarrow u^{(j)}.
\end{aligned}
\tag{3.6}
$$

The proof will be done by induction. Consider first the case $i = j - 1$. Since $\bar{R}^{(j-1)} = R^{(j-1)}$ and the first expression for $u^{(1)}$ in (3.5) equals the first term in (3.6), our assertion (3.7) clearly holds.

Now assume that (3.7) holds for i; we will show that it then also holds for $i - 1$. Starting from (3.6)

$$
\begin{aligned}
r^{(i-1)} &= R^{(i-1)}(r^{(i)} - A^{(i)}u^{(i)}), \\
&= R^{(i-1)}(\bar{R}^{(i)}(r - Au^{(j-i)}) - A^{(i)}u^{(i)}), \\
&= \bar{R}^{(i-1)}(r - Au^{(j-i)}) - R^{(i-1)}A^{(i)}u^{(i)},
\end{aligned}
$$

where we have used relation (3.7). From (3.5) using the superscript $j - i$ instead of i we have the relation

$$
u^{(j-i+1)} = u^{(j-i)} + \sum_{k=1}^{N^{(i)}} \bar{R}_k^{(i)^T} A_k^{(i)^{-1}} \bar{R}_k^{(i)} (r - Au^{(j-i)}).
$$

Using this expression for $u^{(j-i)}$ and factoring $\bar{R}_k^{(i)} = R_k^{(i)} \bar{R}^{(i)}$ in order to pull the factor not depending on k outside the sum results in

$$
\begin{aligned}
r^{(i-1)} = {} & \bar{R}^{(i-1)}(r - Au^{(j-i+1)}) \\
& + \bar{R}^{(i-1)} A \bar{R}^{(i)^T} \sum_{k=1}^{N^{(i)}} R_k^{(i)^T} A_k^{(i)^{-1}} R_k^{(i)} \bar{R}^{(i)} (r - Au^{(j-i)}) - R^{(i-1)} A^{(i)} u^{(i)}.
\end{aligned}
$$

Algorithm 3.2.2: Symmetrized Multiplicative Multilevel Schwarz Preconditioner

$$
\begin{aligned}
u^{(j)} &\leftarrow \sum_{k=1}^{N^{(j)}} R_k^{(j)^T} A_k^{(j)^{-1}} R_k^{(j)} r \\
r^{(j-1)} &\leftarrow R^{(j-1)}(r - Au^{(j)}) \\
u^{(j-1)} &\leftarrow \sum_{k=1}^{N^{(j-1)}} R_k^{(j-1)^T} A_k^{(j-1)^{-1}} R_k^{(j-1)} r^{(j-1)} \\
r^{(j-2)} &\leftarrow R^{(j-2)}(r^{(j-1)} - A^{(j-1)} u^{(j-1)}) \\
&\dots \\
u^{(0)} &\leftarrow \sum_{k=1}^{N^{(0)}} R_k^{(0)^T} A_k^{(0)^{-1}} R_k^{(0)} r^{(0)} \\
u^{(1)} &\leftarrow u^{(1)} + R^{(0)^T} u^{(0)} \\
u^{(1)} &\leftarrow u^{(1)} + \sum_{k=1}^{N^{(1)}} R_k^{(1)^T} A_k^{(1)^{-1}} R_k^{(1)} (r^{(1)} - A^{(1)} u^{(1)}) \\
&\dots \\
u^{(j)} &\leftarrow u^{(j)} + R^{(j-1)^T} u^{(j-1)} \\
u^{(j)} &\leftarrow u^{(j)} + \sum_{k=1}^{N^{(j)}} R_k^{(j)^T} A_k^{(j)^{-1}} R_k^{(j)} (r - Au^{(j)}) \\
u &\leftarrow u^{(j)}.
\end{aligned}
$$

Observe that $\bar{R}^{(i-1)} A \bar{R}^{(i)^T} = R^{(i-1)} A^{(i)}$ by recursively using the relation $A^{(l-1)} = R^{(l-1)} A^{(l)} R^{(l-1)^T}$ for $l = j, j-1, \dots, i$, where $A = A^{(j)}$. Using this and our induction hypothesis on the last factor in the sum results in

$$
r^{(i-1)} = \bar{R}^{(i-1)}(r - Au^{(j-i+1)}) + R^{(i-1)} A^{(i)} \left(\sum_{k=1}^{N^{(i)}} R_k^{(i)^T} A_k^{(i)^{-1}} R_k^{(i)} r^{(i)} - u^{(i)} \right).
$$

Recalling (3.6), we see that the last term is zero and

$$
r^{(i-1)} = \bar{R}^{(i-1)}(r - Au^{(j-(i-1))}),
$$

completing the induction.

The recursive multilevel method given in Algorithm 3.2.7 also has a nonrecursive version. These two methods are also identical in exact arithmetic, with the proof very similar to the one given here.

Algorithm 3.2.3: Multiplicative Multilevel Diagonal Scaling

$$
\begin{aligned}
u^{(j)} &\leftarrow D^{(j)^{-1}} r \\
r^{(j-1)} &\leftarrow R^{(j-1)}(r - Au^{(j)}) \\
u^{(j-1)} &\leftarrow D^{(j-1)^{-1}} r^{(j-1)} \\
r^{(j-2)} &\leftarrow R^{(j-2)}(r^{(j-1)} - A^{(j-1)} u^{(j-1)}) \\
&\cdots \\
u^{(0)} &\leftarrow A^{(0)^{-1}} r^{(0)} \\
u^{(1)} &\leftarrow u^{(1)} + R^{(0)^T} u^{(0)} \\
&\cdots \\
u^{(j)} &\leftarrow u^{(j)} + R^{(j-1)^T} u^{(j-1)} \\
u &\leftarrow u^{(j)}.
\end{aligned}
$$

Algorithm 3.2.4: Recursive Multiplicative Multilevel Diagonal Scaling

- Define $u^{(i)} = \text{V-cycle}(r^{(i)}, i, u^{(i)})$ by
- If ($i == 0$) then

  $u^{(0)} \leftarrow u^{(0)} + A^{(0)^{-1}} r^{(0)}$
- Else

  $u^{(i)} \leftarrow u^{(i)} + D^{(i)^{-1}} r^{(i)}$

  $r^{(i-1)} \leftarrow R^{(i-1)}(r^{(i)} - A^{(i)} u^{(i)})$

  $u^{(i)} \leftarrow u^{(i)} + R^{(i-1)^T} \text{V-cycle}(r^{(i-1)}, i-1, 0)$
- EndIf.

3.2.2 Multiplicative Multilevel Diagonal Scaling

Corresponding to the additive multilevel diagonal scaling method is a **multiplicative multilevel diagonal scaling** algorithm (Algorithm 3.2.3). In this algorithm we traverse sequentially through the levels, but simply do a diagonal scaling on each level.

In this version there is one presmoothing step (the application of $D^{(j-1)^{-1}}$), and no postsmoothing. The prefixes pre and post indicate whether the smoothing is applied before or after the coarse grid correction. It is often useful to express these algorithms by recursion. We can rewrite Algorithm 3.2.3 in the form of Algorithm 3.2.4.

The symmetrized version of this algorithm, Algorithm 3.2.5, is known in the multigrid literature as **V-cycle multigrid,** with one pre- and one postsmoothing step of Jacobi

Algorithm 3.2.5: V-cycle Multigrid with Jacobi Smoothing

- Define $u^{(i)} = \text{V-cycle}(r^{(i)}, i, u^{(i)})$ by
- If ($i == 0$) then

  $u^{(0)} \leftarrow u^{(0)} + A^{(0)^{-1}} r^{(0)}$
- Else

  $u^{(i)} \leftarrow u^{(i)} + D^{(i)^{-1}} r^{(i)}$

  $r^{(i-1)} \leftarrow R^{(i-1)}(r^{(i)} - A^{(i)} u^{(i)})$

  $u^{(i)} \leftarrow u^{(i)} + R^{(i-1)^T} \text{V-cycle}(r^{(i-1)}, i-1, 0)$

  $r^{(i)} \leftarrow r^{(i)} - A^{(i)} u^{(i)}$

  $u^{(i)} \leftarrow u^{(i)} + D^{(i)^{-1}} r^{(i)}$
- EndIf.

relaxation. Again, one is free to use several steps of Jacobi on each level or perhaps a different number of smoothing steps on different levels. Generally, if this algorithm is used without a Krylov subspace accelerator, the Jacobi iteration must be damped in order to ensure convergence.

A variant of this method is the **W-cycle**. In that algorithm the recursive call to the next coarser level is carried out twice in succession. For instance, one possible variant is given by Algorithm 3.2.6. In this example there is one presmoothing and no postsmoothing.

Computational Results 3.2.1: Convergence for Classical Multigrid

Purpose: Compare the convergence behavior of additive and multiplicative diagonal scaling.

PDE: The Poisson equation,

$$-\Delta u = xe^y \quad \text{in } \Omega,$$
$$u = -xe^y \quad \text{on } \partial\Omega,$$

with Dirichlet boundary conditions.

Domain: Unit square.

Discretization: Piecewise linear finite elements.

Calculations: The results for both V- and W-cycles are presented below. In both cases, a single presmoothing and postsmoothing were applied. The table shows

Algorithm 3.2.6: W-cycle Multigrid with Jacobi Smoothing

- Define $u^{(i)}$ = W-cycle($r^{(i)}$, i, $u^{(i)}$) by
- If (i == 0) then

$$u^{(0)} \leftarrow u^{(0)} + A^{(0)^{-1}} r^{(0)}$$

- Else

$$u^{(i)} \leftarrow u^{(i)} + D^{(i)^{-1}} r^{(i)}$$
$$r^{(i-1)} \leftarrow R^{(i-1)}(r^{(i)} - A^{(i)} u^{(i)})$$
$$u^{(i)} \leftarrow u^{(i)} + R^{(i-1)^T}\text{W-cycle}(r^{(i-1)}, i-1, 0)$$
$$u^{(i)} \leftarrow u^{(i)} + D^{(i)^{-1}}(r^{(i)} - A^{(i)} u^{(i)})$$
$$r^{(i-1)} \leftarrow R^{(i-1)}(r^{(i)} - A^{(i)} u^{(i)})$$
$$u^{(i)} \leftarrow u^{(i)} + R^{(i-1)^T}\text{W-cycle}(r^{(i-1)}, i-1, 0)$$

- EndIf.

the convergence behavior as the size of the finest grid is held constant, adjusting the number of levels (and consequently the size of the coarse grid) for two sizes. In all cases, the Krylov subspace accelerator used was GMRES, with a restart of 10. The coarse grid problem was solved exactly, using a direct sparse solver.

Iteration Counts for 10^{-2} *Reduction in Error*

		Iterations		
N_{fine}	Levels	Additive	V-cycle Multiplicative	W-cycle Multiplicative
33	2	7	3	2
33	3	8	3	2
33	4	14	4	2
65	2	7	3	2
65	3	8	3	2
65	4	14	4	2
65	5	22	4	2

Discussion: These results illustrate the excellent behavior of multigrid when solving the Poisson problem with Dirichlet boundary conditions. Note that the V-cycle, while requiring more iterations than the W-cycle, is less expensive to apply per iteration. As expected, the multiplicative method requires fewer iterations than the additive method, substantially fewer in this example.

Algorithm 3.2.7: Pure Multiplicative Multilevel Preconditioner

- $r^{(j)} \leftarrow r, \qquad u^{(j)} \leftarrow u$
- For $i = j, \ldots, 1$

  For $k = 1, \ldots,$ Number of colors $N_c^{(i)}$, on level i

  $$u^{(i)} \leftarrow u^{(i)} + \sum_{l \in \mathrm{Color}^{(i)}_{(k)}} R_l^{(i)^T} A_l^{(i)^{-1}} R_l^{(i)} (r^{(i)} - A^{(i)} u^{(i)})$$

  EndFor

  $r^{(i-1)} \leftarrow R^{(i-1)} (r^{(i)} - A^{(i)} u^{(i)})$

  $u^{(i-1)} = 0$
- EndFor
- For $k = 1, \ldots,$ Number of colors $N_c^{(0)}$, on level 0

  $$u^{(0)} \leftarrow u^{(0)} + \sum_{l \in \mathrm{Color}^{(0)}_{(k)}} R_l^{(0)^T} A_l^{(0)^{-1}} R_l^{(0)} (r^{(0)} - A^{(0)} u^{(0)})$$
- EndFor
- For $i = 1, \ldots, j$

  $u^{(i)} \leftarrow u^{(i)} + R^{(i-1)^T} u^{(i-1)}$
- EndFor
- $u \leftarrow u^{(j)}$.

3.2.3 Multiplicative between Levels, Multiplicative in Levels

In order to obtain a faster numerical convergence rate, though not necessarily faster convergence (in wall clock time) on a parallel computer, we may apply the corrections on each level in a sequential manner. As in Chapter 1, first color the domains $\Omega_k^{(i)}$ on each level i. The multiplicative multilevel Schwarz with coloring preconditioner is given by Algorithm 3.2.7.

By letting each domain $\Omega_k^{(i)}$ have its own color we obtain a purely sequential algorithm. Corresponding to the multilevel diagonal scaling method is the multilevel Gauss-Seidel preconditioner, which is the standard V-cycle multigrid with Gauss-Seidel smoothing. W-cycle variants are also possible.

Computational Results 3.2.2: V-cycle and W-cycle with Gauss-Seidel Smoothing

Purpose: Demonstrate the convergence behavior of multigrid preconditioning with Gauss-Seidel smoothing and both V- and W-cycles.

PDE: The Poisson equation,

$$-\Delta u = xe^y \quad \text{in } \Omega,$$
$$u = -xe^y \quad \text{on } \partial\Omega,$$

with Dirichlet boundary conditions.

Domain: Unit square.

Discretization: Piecewise linear finite elements.

Calculations: The results for both V- and W-cycles are presented below. In both cases, a single Gauss-Seidel presmoothing and postsmoothing was applied. The table shows the convergence behavior as the size of the finest grid is held constant, adjusting the number of levels (and consequently the size of the coarse grid) for two sizes. In all cases, the Krylov subspace accelerator used was GMRES, with a restart of 10. The coarse grid problem was solved exactly, by using a direct sparse solver.

Iteration Counts for 10^{-2} Reduction in Error

		Iterations	
N_{fine}	Levels	V-cycle	W-cycle
33	2	2	1
33	3	3	1
33	4	3	1
65	2	2	1
65	3	2	1
65	4	3	1
65	5	2	1

Discussion: These results show the excellent behavior of multigrid for the Poisson problem. As expected, the W-cycle takes fewer iterations than the V-cycle. The reader should not interpret these results to mean that a simple V-cycle or W-cycle multigrid algorithm is appropriate for all classes of problems. As one moves away from a simple Poisson problem on a structured grid, the convergence behavior of these algorithms can change drastically, requiring often sophisticated changes to the methods to maintain acceptable convergence rates.

Notes and References for Section 3.2

- The hierarchical basis multigrid method (not introduced in the text) is a sequential version of the hierarchical basis method. It may be interpreted as V-cycle multigrid where the smoothing is done only on the grid points that are not present on any of the coarser grids. The hierarchical

Algorithm 3.2.8: Full Multigrid with V-cycle

$$
\begin{aligned}
r^{(j)} &\leftarrow r \\
r^{(j-1)} &\leftarrow R^{(j-1)} r^{(j)} \\
&\cdots \\
r^{(0)} &\leftarrow R^{(0)} r^{(1)} \\
u^{(0)} &\leftarrow A^{(0)^{-1}} r^{(0)} \\
u^{(1)} &\leftarrow R^{(0)^T} u^{(0)} \\
u^{(1)} &\leftarrow u^{(1)} + \text{V-cycle}(r^{(1)}, 1, u^{(1)}) \\
u^{(2)} &\leftarrow R^{(1)^T} u^{(1)} \\
u^{(2)} &\leftarrow u^{(2)} + \text{V-cycle}(r^{(2)}, 2, u^{(2)}) \\
&\cdots \\
u &\leftarrow u^{(j)}.
\end{aligned}
$$

basis multigrid method was introduced in [Ban88] by Bank, Dupont, and Yserentant. A major advantage of this approach is that when new levels of refinement that contain only a few nodes are introduced, the smoothing on the new levels applies only to the new nodes. If a classical V-cycle were applied, the work per cycle could grow as the number of levels times the number of nodes on the finest grid, thus resulting in a far from optimal algorithm.

- The hierarchical basis multigrid preconditioner is the core preconditioner in the well known PLTMG software of Bank [Ban90]. This code is quite general but is only for problems in two dimensions.
- An early multilevel method was proposed in the Soviet Union by Fedorenko in [Fed61]. An analysis of this method was given in Bakhvalov [Bak66].
- The relationship between recursive nested multigrid and a particular multiplicative multilevel method was observed in McCormick and Ruge [McC86]. It is clearly stated in Bramble, Pasciak, Wang and Xu [Bra91a]; generalizations may be found in Xu [Xu92].

3.3 Full Multigrid

The astute reader may have noticed that we have made no indication of how one may generate a "good" initial guess for the multilevel/multigrid iteration. There is a simple, yet very powerful way to do so by solving problems on the coarser grids. First restrict the right hand side of the linear systems down through all the levels to the coarsest grid. Solve the coarsest grid problem. The interpolation of its solution to the next grid provides a good initial guess for that level. We then solve on that level by using a V-cycle, or any other multilevel solver. The approximate solution is then interpolated to the next level to serve as the initial guess. The procedure is repeated until one reaches the finest level. This procedure is referred to as **full multigrid**. The code to calculate $u = Br$ is given in Algorithm 3.3.1.

In the classical multigrid literature, full multigrid is presented slightly differently than in Algorithm 3.3.1. The right hand side for all the levels is calculated directly from the PDE instead of by taking the restriction from the finer grids. This approach does not allow full multigrid to be used as a preconditioner. It is, however, an excellent idea when one is using adaptive mesh refinement and the coarse grid problems must be solved before the fine grid meshes have even been generated.

Computational Results 3.3.1: Convergence Behavior for Full Multigrid

Purpose: Demonstrate the convergence behavior of full multigrid.

PDE: The Poisson equation,

$$\begin{aligned} -\Delta u &= xe^y \quad \text{in } \Omega, \\ u &= -xe^y \quad \text{on } \partial\Omega, \end{aligned}$$

with Dirichlet boundary conditions.

Domain: Unit square.

Discretization: Piecewise linear finite elements.

Calculations: Full multigrid.

Iteration Counts for 10^{-2} Reduction in Error

		Iterations	
N_{fine}	Levels	V-cycle	W-cycle
33	2	2	1
33	3	2	1
33	4	2	1
65	2	2	1
65	3	2	1
65	4	2	1
65	5	2	1

Discussion: These results show the excellent convergence behavior of the full multigrid method applied to the Poisson problem.

Table 3.1. *Algorithms Introduced in this Chapter*

Ordering of Subspaces	Examples
Purely additive	Multilevel diagonal scaling BPX Hierarchical basis method
Multiplicative through levels, additive within the level	V-cycle multigrid with Jacobi smoothing
Additive between levels, multiplicative within the level	Multilevel Gauss-Seidel
Purely multiplicative	V-cycle multigrid with Gauss-Seidel smoothing Hierarchical basis multigrid method
Full multigrid	

Notes and References for Section 3.3

- Full multigrid is based on an earlier idea called nested iteration. Nested iteration is described by Kronsjö and Dahlquist in [Kro71].
- Another class of algorithms related to full multigrid and hierarchical grids is the Cascade method developed by Deuflhard, Leinen, and Yserentant [Deu89]. In this method the solution of elliptic PDEs is computed on a sequence of refined finite element grids (regarded as an outer iteration) starting on a coarse grid. The coarsest grid is solved exactly; refinement of the grid based on local error estimates is introduced. The solution at a new level of refinement is obtained by using, for example, the conjugate gradient method (the inner iteration), starting with the approximation to the solution from the previous coarser grid. In a recent report Deuflhard shows that efficient methods can be obtained without the need for preconditioning of the inner iteration [Deu95]. This approach is used in the software package KASKADE [Roi89a], [Roi89b].

3.4 Practical Multilevel Methods

We have seen how the overlapping Schwarz methods can be extended to general multilevel methods. This derivation has also led us to recognize particular multigrid methods that appear as special cases of a large family of algorithms with almost countless possible variants. The specific algorithms introduced in this chapter are summarized in Table 3.1.

The multigrid literature and the successful multigrid methods that have been used in a wide range of practical applications contain a large variation of algorithms, many specially tailored to the problem at hand and some that would not directly follow from our line of development. We refer the reader to the references at the end of this section to find good starting points to this vast literature.

We will limit our discussion here to a few observations. The large number of multigrid methods is partly due to a great flexibility in choosing various components of the algorithm. Traditionally one can experiment with different smoothers (relaxation algorithms), different restriction and interpolation operators, and different coarse grid strategies. These choices are often guided by basic insights into the properties of the particular PDE at hand. Even in the case of a single, scalar equation with strong anisotropy, changes to the simplest multigrid algorithm are necessary by, for example, semicoarsening or by changing the smoothing procedure from standard Gauss-Seidel to more elaborate alternating line relaxation procedures. This flexibility has been a strength, in the sense of developing very efficient methods for a very large and diverse class of problems, but also a weakness since it has tended to promote the design of special algorithms and codes for every problem. There has been relatively little reuse of software and with a few exceptions limited success in the development of software libraries based on multigrid principles.

Algebraic Multigrid (AMG) represents an effort to effectively limit the possible variations. In this method the choice of interpolation is governed by the relative sizes of the coefficients in the discrete matrix, the algorithm is essentially fixed, using the transpose operator for restriction and the Galerkin approach to generate the coarse problems, while staying with point Gauss-Seidel as the smoother. This approach has proved quite effective on scalar equations, even when anisotropy is present in the coefficients. However, the method takes no advantage of the underlying PDE and tends to run into difficulties when applied to multicomponent systems of PDEs. This is an example of a so-called black box algorithm which ignores all information that is not explicitly available from the discrete matrix.

One way to contrast domain decomposition algorithms with the issues related to multigrid methods, as discussed here, is to regard them as multilevel methods employing potentially a more robust, localized smoother in the form of overlapping subdomain solves. This feature may facilitate parallel implementations and localized data access (see Section 3.6), but also contributes to more modular software design. Clearly, domain decomposition methods have the same potential for problem specific tuning and optimization as multigrid methods. It is also most likely the case that a highly tuned multigrid method (or similarly a domain decomposition method) will provide the most efficient code for any given problem. However, this efficiency often carries a considerable price in development and software maintenance costs.

Alternatively, by focusing on the division of the domain of the PDE into subdomains one may try to build methods that capture enough information from the PDE to provide acceptable efficiency while maintaining robustness. The local solves can be exact or inexact and the domain partitioning may reflect knowledge of the underlying problem to be modeled. When these methods are used as preconditioners in a Krylov subspace based iteration one often finds quite robust convergence with the same coarse grid scheme over a large class of problems. The local problems resolve the structure of the solution while the coarse grid supplies the necessary global coupling to ensure rapid overall convergence. The use of nontrivial overlapping subdomain solvers as part of a preconditioner makes the overall method less sensitive to small variations in the choice of individual components in the algorithm.

Domain decomposition algorithms may provide one way to design building blocks for PDEs that, at the expense of a tolerable increase in computing time, will be amenable to more software reuse. The experience gained so far is extremely limited, and only future developments will establish the potential and benefits of domain decomposition algorithms in this context.

Notes and References for Section 3.4

- Brandt realized very early the enormous potential of multigrid methods. He became their major proponent and pushed for the development of multigrid methods from the 1970s [Bra72], [Bra77].
- A book that serves as an excellent elementary introduction to multigrid methods, explaining the basic principles of why these methods work, is the tutorial by Briggs [Bri87].
- We do not cite the original references to work on multigrid; instead, we refer the reader to the many books on the subject that include Hackbusch [Hac85], McCormick [McC87], [McC89], Wesseling [Wes92], Bramble [Bra93], Griebel [Gri94b], Oswald [Osw94] and Rüde [Rüd93]. The book by Bramble contains perhaps the most extensive and up-to-date presentation of the convergence theory for multilevel methods.
- Some development of the theory of multilevel methods, as of 1989, may be found in Xu [Xu89].
- The book [McC87] contains a fairly complete bibliography (more than 600 references) on multigrid literature up to 1987.
- The early development of Algebraic Multigrid (AMG) is due to Ruge and Stüben; a good reference showing both successes and problems is their chapter in [McC87].
- PETSc (see Appendix 2) provides some object oriented multigrid/multilevel software that is extremely flexible and can handle many of the algorithms introduced above.
- MG-Net, `ftp://casper.cs.yale.edu` and `http://na.cs.yale.edu/mgnet/www/mgnet.html`, maintained by C. Douglas, contains a large database on multigrid and related publications.

3.5 Multilevel Methods as Classical Jacobi and Gauss-Seidel

Some beautiful work has been done on multilevel methods that relates them to standard Jacobi and Gauss-Seidel iterative methods for an extended **semidefinite** linear system.

For simplicity consider the case of nested triangulations. Introduce auxiliary degrees of freedom associated with each node on each of the coarser grids, $u^{(i)}$. Then any element on the finest grid (the jth in this case) may be expressed nonuniquely as

$$u = u^{(j)} + \sum_{i=0}^{j-1} \bar{R}^{(i)^T} u^{(i)}. \tag{3.8}$$

The linear system

$$Au = f$$

may be written as

$$v^T Au = v^T f \qquad \forall v = v^{(j)} + \sum_{i=0}^{j-1} \bar{R}^{(i)^T} v^{(i)}.$$

If one simply inserts into v and u their values in terms of $v^{(i)}$ and $u^{(i)}$, then in the two level case, one obtains the semidefinite linear system,

$$\begin{pmatrix} A & A\bar{R}^{(0)^T} \\ \bar{R}^{(0)}A & \bar{R}^{(0)}A\bar{R}^{(0)^T} \end{pmatrix} \begin{pmatrix} u^{(1)} \\ u^{(0)} \end{pmatrix} = \begin{pmatrix} f \\ \bar{R}^{(0)}f \end{pmatrix}$$

or, equivalently,

$$\begin{pmatrix} A & AR^{(0)^T} \\ R^{(0)}A & A^{(0)} \end{pmatrix} \begin{pmatrix} u^{(1)} \\ u^{(0)} \end{pmatrix} = \begin{pmatrix} f \\ R^{(0)}f \end{pmatrix}. \tag{3.9}$$

In the case of three levels the resulting linear system is

$$\begin{pmatrix} A & A\bar{R}^{(1)^T} & A\bar{R}^{(0)^T} \\ \bar{R}^{(1)}A & \bar{R}^{(1)}A\bar{R}^{(1)^T} & \bar{R}^{(1)}A\bar{R}^{(0)^T} \\ \bar{R}^{(0)}A & \bar{R}^{(0)}A\bar{R}^{(1)^T} & \bar{R}^{(0)}A\bar{R}^{(0)^T} \end{pmatrix} \begin{pmatrix} u^{(2)} \\ u^{(1)} \\ u^{(0)} \end{pmatrix} = \begin{pmatrix} f \\ \bar{R}^{(1)}f \\ \bar{R}^{(0)}f \end{pmatrix}$$

or, equivalently,

$$\begin{pmatrix} A & AR^{(1)^T} & AR^{(1)^T}R^{(0)^T} \\ R^{(1)}A & A^{(1)} & A^{(1)}R^{(0)^T} \\ R^{(0)}R^{(1)}A & R^{(0)}A^{(1)} & A^{(0)} \end{pmatrix} \begin{pmatrix} u^{(2)} \\ u^{(1)} \\ u^{(0)} \end{pmatrix} = \begin{pmatrix} f \\ R^{(1)}f \\ R^{(0)}R^{(1)}f \end{pmatrix}.$$

Here we have used the fact that $A^{(j-1)} = R^{(j-1)}A^{(j)}R^{(j-1)^T}$ and $\bar{R}^{(i)} = \Pi_{k=i}^{j-1} R^{(k)}$. Note that in the case where A is symmetric the resulting larger matrices are also symmetric.

The key observation is that the various multilevel and multigrid algorithms given in the previous sections are simply Jacobi and Gauss-Seidel applied to the above extended linear systems. To demonstrate this we consider, for simplicity, a two level method. If we use a partial block Jacobi preconditioner for the linear system (3.9) we obtain the Richardson scheme,

$$\begin{pmatrix} u^{(1)^{n+1}} \\ u^{(0)^{n+1}} \end{pmatrix} \leftarrow \begin{pmatrix} u^{(1)^n} \\ u^{(0)^n} \end{pmatrix} + \begin{pmatrix} D^{-1} & 0 \\ 0 & A^{(0)^{-1}} \end{pmatrix} \left[\begin{pmatrix} f \\ R^{(0)}f \end{pmatrix} - \begin{pmatrix} A & AR^{(0)^T} \\ R^{(0)}A & A^{(0)} \end{pmatrix} \begin{pmatrix} u^{(1)^n} \\ u^{(0)^n} \end{pmatrix} \right].$$

Using (3.8) we know that $u^{n+1} = u^{(1)^{n+1}} + R^{(0)^T} u^{(0)^{n+1}}$. We can therefore multiply the second component, that is, $u^{(0)}$, of the vectors in the above Richardson scheme by $R^{(0)^T}$ in order to obtain an iteration scheme for u^{n+1},

$$u^{n+1} \leftarrow u^n + (D^{-1} + R^{(0)^T} A^{(0)^{-1}} R^{(0)})(f - Au^n).$$

But this is exactly what is obtained by the two level diagonal scaling method, defined in Algorithm 3.1.2 with $j = 1$. Similar calculations can be done for the multiplicative methods.

This idea may be extended to a wide variety of domain decomposition and multilevel algorithms. Let R_i denote the generic restriction matrices and $C_{i,i=1,\dots,p}$, denote the subproblems associated with R_i. In the Galerkin case $C_i = R_i A R_i^T$. Consider the rectangular matrix

$$R = \begin{pmatrix} R_p \\ R_{p-1} \\ \dots \\ R_1 \end{pmatrix}$$

and the block diagonal matrix

$$D = \begin{pmatrix} C_p & & & \\ & C_{p-1} & & \\ & & \dots & \\ & & & C_1 \end{pmatrix}.$$

Then a pure additive preconditioner may be written as

$$R^T D^{-1} R.$$

This can be interpreted as a change (and expansion) of basis, followed by a block diagonal scaling and a change back to the original basis. See also the conclusion of Section 3.1.3.

The remarkable fact is that the multiplicative algorithms may be put in the same framework. Let L denote the lower triangular matrix whose elements are given by

$$L_{ij} = R_i A R_j^T.$$

Similarly U is the upper triangular matrix whose entries are given by

$$U_{ij} = R_j A R_i^T.$$

The preconditioner for the multiplicative algorithm that marches down from R_p to R_1 can be written as

$$R^T (L + D)^{-1} R.$$

The symmetrized version that first traverses down through the subspaces and then back up is given by

$$R^T (U + D)^{-1} D (L + D)^{-1} R.$$

Notes and References for Section 3.5

- The construction of the semidefinite system is due to Griebel [Gri94a], [Gri94b] and Griebel and Oswald [Gri95]. This interpretation, which is related to product spaces created from the Schwarz subspaces, is useful in analysis of the algorithms, especially for the multiplicative methods.

3.6 Complexity Issues

In order to understand the issues in choosing the number of levels to use, we need to model the computational complexity of the various parts of the algorithm. In this section, we will present a simple complexity model and discuss some of the implications for domain decomposition. This section complements the discussion in Section 2.6.2 on solving the coarse problem in parallel but can be read independently.

3.6.1 A Simple Model

The simplest model of complexity counts only the floating point operations and gives them all the same weight (e.g., an addition takes the same amount of time as a division). Using this model, we can compare the cost of two algorithms by simply counting the floating point operations.

As a simple example, consider a Poisson problem on a square, discretized with an $n \times n$ mesh. Using a banded direct solver (Gaussian elimination) requires roughly n^4 floating point operations. If a two level Schwarz method with an $m \times m$ decomposition of the domain is used, the complexity has the following contributions:

- each subdomain: $(n/m)^4$,
- coarse grid problem: m^4,
- interpolation and restriction: n^2,
- vector operations in the Krylov subspace method: n^2.

The cost for a single iteration is roughly

$$T = m^2 \left(\frac{n}{m}\right)^4 + m^4 + n^2.$$

The cost is minimized for

$$m = \frac{n^{2/3}}{2^{1/6}}$$

and is roughly $T = n^{8/3}$. If the number of iterations required is independent of m, then this gives a way of choosing the number of subdomains. Assuming that this is the case, our result shows that the overall complexity of a domain decomposition algorithm is lower than the complexity of the solver applied to the original problem. Note that for a fixed problem size n, the cost of solution may not be smaller, because of a possibly much larger coefficient of the leading term (mainly due to the iteration count), but for large n, we will be ahead. Moreover, this way of organizing the solution procedure may also lend itself to better data locality and parallel implementations.

Note that an optimal method (defined as performing a constant amount of work for each degree of freedom) would require $T_{opt} = n^2$ time. We are already fairly close to that, but we can do better by applying the algorithm to both the "coarse grid" problem and the subdomain problems, as discussed in this chapter.

Note that if an optimal method exists for both the coarse grid and the subdomain problems, then the domain decomposition algorithm is also optimal, since the total time is the sum of m^2 subdomains with work $(n/m)^2$ and the coarse grid problem of size $m^2 < n^2$. Thus, given any optimal method for the subproblems, we can make domain decomposition methods optimal as well.

The choice of a direct banded method was made in the above analysis in order to simplify the discussion. In practice, even when choosing a direct method, it may be better to use a sparse technique that has a lower time complexity. Obviously, given the complexity of the subdomain solver and of the coarse grid problem, it is a relatively simple matter to estimate the relationship between the sizes of m and n in order to minimize the overall complexity of the computation.

An extreme case is to keep the subdomain size n/m fixed. As the overall problem size increases we then get a large coarse grid problem. In order to keep the complexity of our algorithm low, we should apply one of the multilevel algorithms described in this chapter to the coarse grid problem. Provided that we can solve this with an optimal (or near optimal) method like multigrid or multilevel domain decomposition, we get an overall method with (nearly) optimal complexity; that is, the work per iteration will be proportional to the number of grid points n, independent of the complexity of the subdomain solver.

In practice, one often uses inexact solvers for both the subdomains and the coarse grid problem. This will lead to an increase in the number of outer Krylov subspace iterations, but may also lead to a substantial reduction in the total computing time. A good algorithm should then balance the accuracy of the subdomain solves with the accuracy of the coarse grid solver as well as finding the right trade-off between inner and outer iterations. Additionally, in most applications, one normally prefers to keep the overlap fixed in terms of grid points instead of geometric overlap of the subdomains. This causes a slight increase in the number of outer iterations as the problem size is scaled up unless we keep the subdomain size fixed and increase the number of subdomains. The alternative will considerably increase the number of grid points shared between two or more subdomains and most often result in an increase of the total computational effort. Both extensive experimental as well as theoretical work indicates that the use of minimal overlap (i.e., typically one or two lines of grid points) tends to minimize the overall execution time of domain decomposition algorithms. A more refined complexity model must take these effects into account, in particular the extra work associated with the overlap area.

3.6.2 Toward Better Models

By taking the factors from the previous discussion into account one can get a reasonably accurate model of the number of floating point operations required in the computation. Unfortunately, on modern computers, this may not provide an effective way of estimating the running time of a program. One obvious reason is that computers with special floating point hardware for vector operations perform scalar operations much more slowly than vector operations. A more important reason is that it often is memory operations (loads and stores) that actually dominate the cost of a computation. This is because main memory is much slower than the CPU. It is more difficult to model the cost of memory operations, because this cost usually depends on the past history of the computation. To prevent the CPU from being starved for data, hardware designers have developed a hierarchical memory structure of increasingly fast (but small) memories. On most workstations (and the nodes of parallel computers), these small fast memories are called **caches**. Data intended for the CPU must (usually) be first copied into a cache. Since the cache is small (relative to our computation), a *replacement strategy* is used to discard data in the cache to make room for new data. Modeling the effect of this replacement strategy on the time to access memory is difficult. However, there are a few rules of thumb for identifying sources of good memory performance on cache-based computers. They include (a) keeping memory references within the cache and (b) avoiding references that replace data in the cache that will be needed soon. One easy way to accomplish these is to operate on small blocks of data that are contiguous in memory. Domain decomposition gives a way to organize the computation in line with these principles.

3.6.3 Complexity for Parallel Computers

Parallel computers can be thought of as computers with a hierarchical memory where the memory on another processor is relatively expensive to access (compared to local memory). On many systems, the cost of accessing remote memory (whether through message-passing or some sort of shared-memory operation) is large enough that the cost must be accounted for in even the simplest complexity models. The simple model introduced in Section 2.6.2 may be used. In this model the cost is $s + rn$, where s is the startup time or latency (the time it takes for the first byte to arrive), n is the number of bytes being moved, and r is the time to send a byte from one processor to another (the reciprocal of r is called the communication bandwidth).

We adopt the Single Program Multiple Data (SPMD) programming model for discussion of domain decomposition algorithms on parallel computers. Our algorithms can then be viewed as implemented by a single computer program. A copy of this program is executing on each processor operating on its own locally stored data and exchanging messages with other processors. The computer consists of independent processors each having access to a local memory and to some communication network in order to exchange data with other processors. An important parameter in such a system is ρ, the ratio of the (average) time it takes to send a (64 bit) word to another processor, to the (average) time the processor requires to multiply two such (floating point) words. This ratio can range from around one on fine grained massively parallel computers to several thousands or more on message-passing machines using state of the art microprocessors interconnected by some dedicated switch or by a local area network. In order for the computer to be reasonably balanced, a high value of ρ is usually compensated by a large local memory per processor (and correspondingly fewer processors for the computer with the same aggregate computational power).

Some of the more important concepts that affect the elapsed time of our domain decomposition algorithms on a parallel computer are

- **Data structure.** The data belonging to one or more subdomains is stored in a local memory. Depending on the interprocessor network topology and its characteristics it may be desirable to map subdomains to processors in a way that minimizes communication distance and/or data contention. The overlapping data is often copied between processors, creating a slight increase in space complexity compared with that of a single processor algorithm. The data structure for the coarser grid problems and the degree of data locality that can be achieved in this phase of the algorithm may critically influence the overall complexity of the method. Our previous discussion of memory hierarchies in single processor systems often applies to each processor in a parallel computer.

- **Communication.** The subdomains on a given level exchange data in the overlap region. This communication is local (i.e., nearest neighbor); this may be of importance for some machines. For a local memory holding k^2 or k^3 grid points (from one or more subdomains) there is a surface to volume effect relating the arithmetic of a local solve to the amount of communication. In the case of an optimal local solver, the ratio of arithmetic to communication will be $O(k)$ for each subdomain; a local solver with higher complexity will make this factor even more favorable. A ratio considerably larger than ρ will help make the overall algorithm compute bound. (i.e., computational time will dominate over communication time). The interlevel exchange of data through interpolation and restriction generally involves longer range and less structured communication. This cost depends very much on how and where the coarser level problem is being solved. The overall contribution to communication complexity can be controlled since the coarser problem only sends a small, fixed number of bytes to each fine grid subdomain. Adding up the interlevel communication for all the coarser levels therefore normally contributes a $O(\log(p))$ communication cost if we have N subdomains distributed to p processors on the finest level.
- **Load balancing.** Achieving good load balancing is often quite significant for overall performance of a parallel algorithm. Allocation of subdomains to processors will invariably result in an uneven distribution of work. In practice, the size of individual subdomains will also show some variation. Additive algorithms are normally easier to load balance, partially because the number of parallel tasks usually is significantly larger than in the multiplicative case. If, in addition, the coarse level problem is solved on a single processor then this adds another source of possible imbalance. Ideally, some form of dynamic scheduling of computational tasks to processors provides the best load balancing, but this may not be feasible since the cost of moving subdomain data between processors may be prohibitive.
- **Synchronization.** A synchronization point in the algorithm is a point where all processors are required to be at the same time. Examples of synchronization points are the computed inner products in the Krylov subspace iteration. Similarly, we get synchronization points between levels in a multiplicative algorithm and when using the recursive form of interpolation/restriction discussed in Section 3.1. Depending on the implementation, there may also be synchronization points for each color in a multiplicative Schwarz code. Having many synchronization points tends to increase load imbalance; one should therefore search for implementations with as few synchronization points as possible.
- **Scalability.** The term scalability tries to express the benefit of solving large problems on large parallel computers. (This is what large machines are designed for; smaller problems should routinely be solved on easy to access workstations.) Traditionally,

a measure of **speedup** has been used to judge the quality of a parallel algorithm running on a parallel computer. Speedup is defined as the time to run a problem of size n (possibly using the best known sequential algorithm) on one processor, divided by the time it takes to run the problem using p processors. Speedup divided by p is called the parallel **efficiency**. The efficiency indicates how well we are using the processors. Unfortunately, the efficiency will approach zero as we increase the number of processors for a fixed problem size n.

The more sensible use of large, parallel computers is to solve larger problems; that is, we should scale the problem with the increased total memory available to us when we use more processors. We will say that an algorithm is scalable if the efficiency is constant (or at least bounded away from zero) when we increase the number of processors p, but at the same time scale the problem size such that the memory usage per processor is fixed. As an example, if we solve a dense linear system of equations of dimension n, then we need n^2 memory to store the problem. If we double the problem size to $2n$, then we need four times as many processors in order to not increase the storage per processor. We would then say that our algorithm was scalable if the running time only increased by a factor of 2.

Another desirable property is **constant time to solution**: that is, if we double the size of our problem and the size of the computer, then our algorithm will take the same time as before. A little reflection will make it clear that this can only happen if we have what we have called an optimal algorithm that, additionally, is scalable. In theory, this is exceedingly difficult to obtain since complexity estimates for most parallel algorithms will contain terms like $\log(p)$ from the calculation of inner products or related to the number of levels in a multilevel algorithm. These terms are often completely irrelevant in practice, since the size of a local memory typically is much larger than p on any machine that is built today (and in the future). In practice, we can design and implement multilevel Schwarz methods that are nearly scalable for many elliptic PDEs. Since the complexity of these methods is (nearly) proportional to the number of grid points, we also achieve (nearly) "constant time to solution" for such problems. An example of this is given in Computational Results 3.6.1.

Computational Results 3.6.1: Multilevel Domain Decomposition

Purpose: Demonstrate scalability and convergence properties of a multilevel algorithm on a massively parallel computer.

PDE:

$$\begin{aligned} -\nabla \cdot k(x, y)\nabla \boldsymbol{u} &= 4 - 2e^x \cos y \qquad &&\text{in } \Omega, \\ \boldsymbol{u} &= x^2 + y^2 - xe^x \cos y &&\text{on } \partial\Omega, \\ k(x, y) &= 0.01 + 100(x^2 + y^2). && \end{aligned}$$

Domain: Unit square.

Discretization: Linear, (P_1) finite elements on a uniform mesh.

Algorithm: Hybrid preconditioner. The two level preconditioner given in (3.1) is used. We keep the subdomain overlap fixed at $2h$. The subdomain problems are solved approximately by using five symmetric Gauss-Seidel (SGS) iterations. The coarse grid problem $A^{(j-1)}$ is approximated with three iterations of Algorithm 3.2.5, modified to use two pre- and two postsmoothing steps at each level.

Calculations: Convergence was declared when the 2-norm of the residual in the conjugate gradient iteration had been reduced by 6 orders of magnitude, $||r_k||_2 \leq 10^{-6}||r_0||_2$.

Scalability, Keeping One Subdomain of Fixed Size per Processor

Processors	Subdomain	Grid points	Iterations	Time
1024	29 × 29	802,816	43	93.8s
4096	29 × 29	3,211,264	43	94.0s
16384	29 × 29	12,845,056	42	92.2s

Fixed Problem Size, with Increasing Number of Processors

Processors	Subdomain	Grid points	Iterations	No. SGS	Time
1024	29 × 29	802,816	43	5	93.8s
4096	15 × 15	802,816	27	5	18.8s
16384	8 × 8	802,816	19	5	5.0s
1024	29 × 29	802,816	26	40	252.2s
4096	15 × 15	802,816	21	15	28.5s

One V-cycle with Two Pre- and Two Postsmoothing Jacobi Relaxation Steps as Approximate Subdomain Solver. (The Function $k(x, y) \equiv 1$; the Subdomain Size is Fixed.)

Processors	Subdomain	Grid points	Iterations	Time
1024	15 × 15	200,704	20	7.9s
4096	15 × 15	802,816	20	8.1s
16384	15 × 15	3,211,264	20	8.2s
4096	7 × 7	147,456	15	1.8s
16384	3 × 3	65,536	16	1.0s

Discussion: In the first table we keep the subdomain size fixed and scale up the problem and the size of the machine by a factor 16. We observe that the time is virtually unchanged. We use a very naive subdomain solver, five iterations of symmetric Gauss-Seidel (SGS). In this case the subdomain solution is quite inaccurate and the number of conjugate gradient iterations is considerably higher than it would be with a more accurate subdomain solver.

In the second table we start out with the same problem on 1024 processors, but keep this problem fixed while increasing the size of the machine. This results in a corresponding decrease in the subdomain size. The number of iterations decreases as the subdomain solutions become relatively more accurate (since we use the same number of symmetric Gauss-Seidel iterations), but also because the coarse grid influence increases (since it is getting refined). In the two last rows of the table we show how the cost increases if we insist on spending more work on the subdomain solves in order to reduce the number of conjugate gradient iterations. The number of outer iterations will increase slightly with the size of the subdomain since we keep the overlap fixed at $2h$.

The third table shows results from a Poisson problem using a single V-cycle (with two pre- and two postsmoothing Jacobi relaxation steps) as an approximate subdomain solver. The first three rows verify that we maintain almost perfect scalability when increasing the problem and the machine size. The two last rows show decreasing subdomain size while increasing the machine size. If we adjust for the variation in grid points we get a speedup around 3.2 when moving from 1024 to 4096 processors, while the computation time per grid point actually increases when we change from 4096 to 16384 processors. This effect was hidden by other effects when we used the nonoptimal symmetric Gauss-Seidel solver in the second table.

Notes and References for Section 3.6

- Skogen [Sko92] carried out extensive numerical tests showing that the lowest overall complexity often was obtained with minimal amount of overlap. Motivated by these observations Dryja and Widlund provided some theoretical explanation for the favorable rate of convergence in [Dry94b]; see also Theorem 2 in Section 5.3.1.
- Computational Results 3.6.1 were performed by using a MasPar MP-2216 computer. The algorithms are due to Bjørstad and Skogen [Bjø92], [Sko92].

3.7 Implementation Issues

On sequential computers, the implementation of multilevel methods, either as preconditioners or in the classical multigrid style, is often an easy modification of a two level code. On parallel computers this is usually not the case, because of the use of special purpose software for calculating the coarse grid restriction and interpolation. This is why we recommend using general purpose parallel sparse matrix routines (like those discussed in Appendix 2) to implement all of the important steps in the iterative process (at least until careful study shows that special purpose routines can provide a significant improvement). Unfortunately, most codes implementing parallel multilevel preconditioners are handcrafted for a particular problem on a particular machine. We hope this will change in the next few years.

The question as to which preconditioner is appropriate for a particular application and computer is often difficult to answer. Our general feeling is that one should select the preconditioner that has the best numerical convergence qualities. The best results will normally be obtained if one can parallelize the method that minimizes the computing time on a sequential computer. In general this is full multigrid using a multiplicative V- or W-cycle. If this is not feasible then the algorithm with the most multiplicative features, that can still be parallelized to utilize the machine fully, may be the best choice. To select a good preconditioner on the basis of these general guidelines is often hard. It is not unusual to find a trade-off between the convergence properties of the sequential algorithms and the work that must be invested to obtain an efficient parallel version of the same methods.

Let us first consider the coarse grid problem. We have seen that we most likely need to implement this solver by using a multilevel algorithm in order to end up with a method that scales well. We have often found that it helpful logically to consider the algorithm as a two level method, even if the second level is itself a multilevel algorithm. One is then free to consider the two stages independently with a possible gain in flexibility and modularity. Advice on how to proceed with the coarse grid problem has been given in Section 2.6.2. The choices made with respect to the coarse grid will determine whether the coarse grid problem will be solved in parallel (in an additive fashion) or sequentially (in a multiplicative fashion) relative to the fine grid problems.

The decision on what type of domain decomposition method to use for the fine grid problem will further select one of the four basic two-level algorithms described in Chapter 2. A choice of data structure and the representation of overlapping subdomains must be made. As indicated in the previous section, the amount of overlap should generally be quite small. Besides the degree of parallelism that is required for the target machine, there are a few other factors that may influence the choice between an additive or a multiplicative fine grid implementation. If the problem at hand is symmetric, then the choice of the conjugate gradients method as the Krylov subspace accelerator will tend to favor an additive method. We have seen that the need to symmetrize the multiplicative algorithms requires a substantial amount of additional work, since we need to traverse the different colored subdomains twice. Alternatively, one can choose to ignore the symmetry and use another Krylov subspace method, but these methods usually require more storage and inner products or two matrix vector products per iteration. The purely additive method is also sometimes quicker to implement since it requires less code.

When the overall structure of the algorithm has been decided, there remain many details that can have substantial impact on the final program. We strongly prefer to implement these methods in an incremental way. First, even though the target computer may be a parallel computer, all development work should be done on a workstation if at all possible. The actual parallelization work should be a small and focused effort toward the end of the project. Second, in a development phase, first solve the coarse grid problem as well as all subdomain problems by using a direct solver. One should separately test the preconditioner and the matrix-vector multiply portions of the code. When the expected convergence behavior of the method (using exact solvers) has been verified, it is time to substitute more advanced modules to increase the speed and scalability of the overall algorithm. At this time the coarse grid solver may be replaced by a multilevel approximation and the subdomain solvers similarly with some more efficient iteration like a multigrid V-cycle.

In terms of software development, the more flexible the software, the more easily a wide variety of multilevel algorithms may be implemented.

3.8 Variational Formulation

In this section, we demonstrate that for multilevel methods, as for one level methods (Section 1.6) and two level methods (Section 2.8), the corrections calculated during the iterative process are projections (or approximate projections) of the error onto subspaces of the fine grid space. For simplicity we restrict attention to nested subspaces.

Again for technical reasons, we must work in the context of Galerkin finite elements. Let $V^j = V$ denote a Lagrangian C_0 finite element space associated with the finest grid. Further assume that $V^{j-1}, \ldots, V^0$ are Lagrangian C_0 finite element spaces associated with the coarser grids of equal or lower polynomial degree. Then the spaces are naturally nested,

$$V^0 \subset V^1 \subset \cdots \subset V^{j-1} \subset V^j.$$

We will use ϕ_l^i to denote the basis functions on level i.

The subspaces on each level are now decomposed in the same way as in Chapter 1. Let

$$V_k^i = \text{span}\{\phi_l^i | \text{supp}(\phi_l^i) \subset \Omega_k^{(i)}\}.$$

Thus $V_k^i \subset V^i$.

The correction $\bar{R}_k^{(i)^T} A_k^{(i)^{-1}} \bar{R}_k^{(i)} A$ is the matrix representation of the projection onto the subspace V_k^i in the inner product induced by A. The proof for this result is identical to that given in Section 2.8 for the two level methods.

In fact, even the traditional recursive, nested multigrid method involves projections of the error onto subspaces of the finest grid; see Section 3.2.1. Thus they may be analyzed by using the same mathematical tools as the domain decomposition methods; see Chapter 5.

4
Substructuring Methods

In the previous three chapters we have discussed domain decomposition methods where the subdomains explicitly overlap. In this chapter we introduce a large class of methods that use nonoverlapping subdomains. Domain decomposition methods with no overlap between the subdomains are called **substructuring methods**. This name has a long tradition in the structural analysis community, but almost always in relation to direct solution algorithms based on explicit computation and factorization of a sequence of Schur complement matrices. Motivated by this approach, active research over the last 15 years has resulted in promising iterative alternatives that are the subject of this chapter. We refer to these algorithms as **iterative substructuring methods** or **Schur complement methods**. The methods of this chapter may be particularly attractive if new and more efficient solution algorithms are needed, while at the same time preserving most of the data structures and concepts already present in existing large engineering software packages.

In Chapter 1 we demonstrated that for overlapping domain decomposition methods the rate of convergence decreases as the overlap is reduced. To overcome this problem the iterative substructuring algorithms have an additional feature not seen in the overlapping methods, the interface solve. In Section 4.3.6 we will introduce a method that merges the features of both the overlapping and the substructuring methods.

A large number of Schur complement methods have been proposed over the last 15 years. A comprehensive discussion cannot be given in a single chapter or even in an entire book. Therefore, in this chapter we survey the main highlights of the more attractive methods.

4.1 Direct Substructuring Methods

The earliest successful, systematic approach to solving elliptic PDEs on computers was the development of computational structural analysis, beginning in the 1950s. The structural engineers developed two equivalent computational procedures, known as the force method and the displacement method. Gradually the displacement method gained more popularity and it evolved into the finite element method as we know it today.

Since the very beginning, direct methods for solving the resulting linear equations have been the exclusive method of choice. The size of the problems being solved has increased to several million unknowns, constantly pushing available computer resources to the limit.

Domain decomposition was adopted early as a systematic way of organizing a large

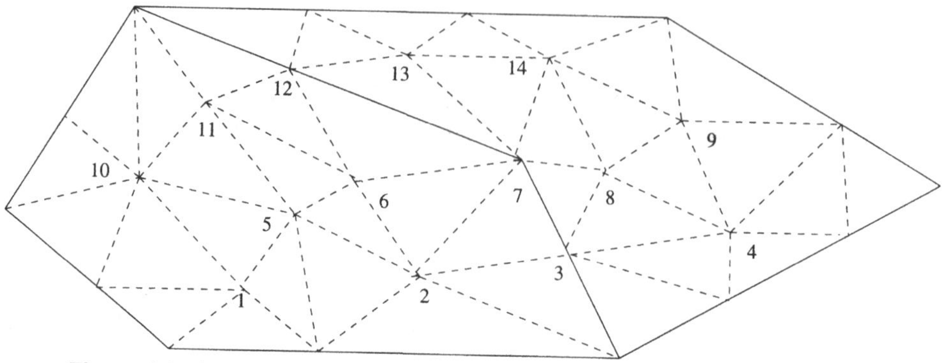

Figure 4.1. Substructuring with Two Subdomains and Natural Node Ordering

structural analysis. Przemieniecki, 1968 [Prz85], explains and motivates the subject in his book *Theory of Matrix Structural Analysis:*

> In applying matrix methods of analysis to large structures, the number of structural elements very often exceeds the capacity of available computer programs, and consequently some form of structural partitioning must be employed. Structural partitioning corresponds to division of the complete structure into a number of substructures, the boundary of which may be specified arbitrarily; however, for convenience it is preferable to make structural partitioning correspond to physical partitioning. If the stiffness or flexibility properties of each substructure are determined, the substructures can be treated as complex structural elements, and the matrix displacement or force methods of structural analysis can be formulated for the partitioned structure. Once the displacements or forces on substructure boundaries have been found, each substructure can then be analyzed separately under known substructure-boundary displacements or forces, depending on whether displacement or force methods of analysis are used.

Substructuring is therefore a way of organizing the direct factorization of large sparse linear systems arising from the discretization of PDEs. An additional feature, not considered by Przemieniecki, is, as we shall see, that substructuring facilitates the design of effective parallel algorithms.

Consider the domain as depicted in Figure 4.1. Assume we have discretized the PDE and obtained the linear system

$$Au = f.$$

In order to introduce some concepts informally, consider the case where our discretization is as indicated in Figure 4.1 and where the boundary nodes have been eliminated from the linear system. If we number the unknowns *naturally* by rows, then the stiffness matrix A has the nonzero structure given in the left side of Figure 4.2. The matrix is sparse and has the typical structure of a two-dimensional discrete problem with natural ordering of the unknowns. In the direct factorization of sparse linear systems, the ordering of the unknowns has a large effect on the required amount of arithmetic work and data storage. The amount of additional storage needed in the factors is referred to as **fill** or **fill-in**. Our example matrix has symmetric nonzero structure, but more generally it may be nonsymmetric, depending on the actual PDE and the discretization scheme.

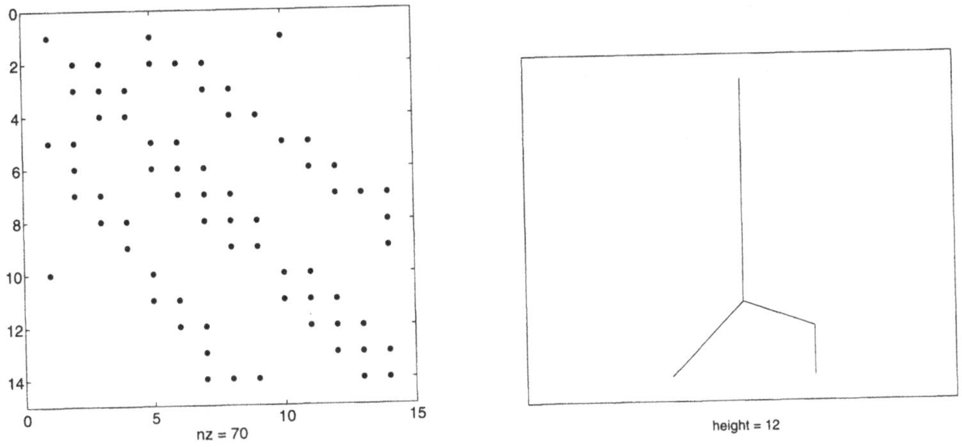

Figure 4.2. Matrix with Natural Ordering and Its Elimination Tree

In order to simplify our presentation, we will limit the discussion to the symmetric, positive definite case; however, the reader should note that the concepts apply also in the more general case.

When the matrix A, of our tiny example, is factored (with natural ordering) it suffers a fill of 28 elements in the triangular factor. (The band fills completely with the exception of the first four rows of the matrix.) The rightmost part of Figure 4.2 shows its **elimination tree**. This graph gives the dependencies between nodes in the elimination process and therefore indicates important potential for parallel work in the factorization. We see that the factorization can start by doing the first node and the next 3×3 block (since there is no coupling between these two blocks) in parallel, but then the remaining part of the factorization must proceed sequentially.

Returning to Przemieniecki and his "division of the complete structure into a number of substructures, the boundary of which may be specified arbitrarily," consider dividing the structure into two parts, each having roughly the same number of nodes or elements (5 and 20 in our example) by ordering the nodes on the *interface* last. These nodes are numbered 3, 7, and 12 in our example from Figure 4.1. We can apply this idea recursively: ordering nodes 5 and 11 last within the first substructure further divides this into two equally sized subsubstructures. Ordering node 14 last in the second substructure has a similar effect, but this time the nodes split into two sets of different size. This procedure can be generalized into an algorithm called **nested dissection**. In Figure 4.3 we see the result applied to our example problem. This ordering suffers a fill of 11, and, in addition, has a much lower elimination tree, suggesting good parallel properties. The nested dissection algorithm can be viewed as a systematic way of recursively dividing a domain into substructures. Conversely, a domain partitioning that corresponds to physical substructures can be seen as providing a nested dissection ordering of the nodes. In engineering practice one normally does multisectioning; that is, the structure is divided into a variable number of smaller substructures at each step of the recursive process. The division is according to physical or modeling partitions

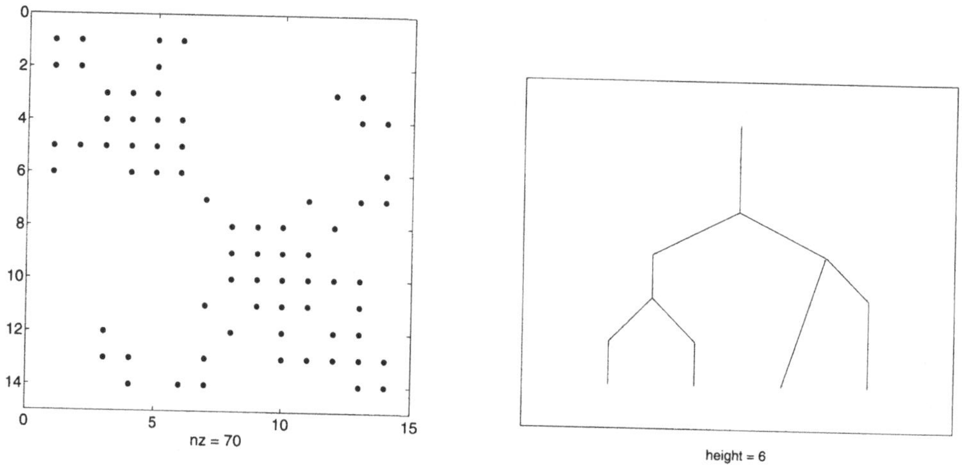

Figure 4.3. Matrix with Nested Dissection Ordering and Its Elimination Tree

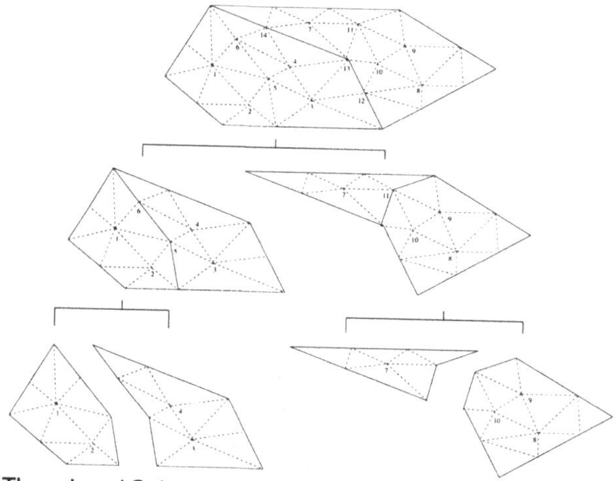

Figure 4.4. The Three Level Substructure Elimination Tree with Nested Dissection Ordering

and the recursion stops when there still may be hundreds or thousands of nodes left in each first level substructure.

Figure 4.4 shows our example divided into substructures according to the ordering given in Figure 4.3. Let $A^{(i)}$ be the contribution to the stiffness matrix A obtained from substructure Ω_i. Following Przemieniecki, we partition $A^{(i)}$ into two blocks, the first block associated with the interior of Ω_i and the second block associated with the boundary interface between Ω_i and the remaining part of the structure. We will use the subscript I to denote portions of the matrix associated with nodes in the *interior* and on the true exterior boundary of any substructure while the subscript B denotes those portions associated with the *interior boundary* of a substructure facing one or more neighboring substructures. For these interface nodes, more than one substructure contributes (from the appropriate $A^{(i)}$) to A. The substructure in question will always be identified with a superscript. In this chapter, we will frequently refer to submatrices and corresponding subvectors that are associated with various geometrically defined subregions such as interior, boundary, edges, faces, and vertex points. The corresponding symbols will consistently be subscripted with capital letters (i.e., I, B, E, F, V, etc.). This partitioning induces a consistent partitioning of the unknowns of each substructure into two sets, the interior degrees of freedom $u_I^{(i)}$ and the boundary variables $u_B^{(i)}$. The stiffness matrix associated with the individual subdomains may be written as

$$A^{(i)} = \begin{pmatrix} A_{II}^{(i)} & A_{IB}^{(i)} \\ A_{BI}^{(i)} & A_{BB}^{(i)} \end{pmatrix}.$$

Combining them one obtains (in the two substructure case) the stiffness matrix

$$A = \begin{pmatrix} A_{II}^{(1)} & 0 & A_{IB}^{(1)} \\ 0 & A_{II}^{(2)} & A_{IB}^{(2)} \\ A_{BI}^{(1)} & A_{BI}^{(2)} & A_{BB}^{(1)} + A_{BB}^{(2)} \end{pmatrix}.$$

We see that if we were to do a direct factorization of A, the blocks $A_{II}^{(1)}$ and $A_{II}^{(2)}$ could be factored independently, in parallel. After this, the blocks $A_{BI}^{(1)}$ and $A_{BI}^{(2)}$ could also be eliminated in parallel.

Let us express A in factored form,

$$\begin{pmatrix} I & 0 & 0 \\ 0 & I & 0 \\ A_{BI}^{(1)} A_{II}^{(1)^{-1}} & A_{BI}^{(2)} A_{II}^{(2)^{-1}} & I \end{pmatrix} \begin{pmatrix} A_{II}^{(1)} & 0 & 0 \\ 0 & A_{II}^{(2)} & 0 \\ 0 & 0 & S^{(1)} + S^{(2)} \end{pmatrix} \begin{pmatrix} I & 0 & A_{II}^{(1)^{-1}} A_{IB}^{(1)} \\ 0 & I & A_{II}^{(2)^{-1}} A_{IB}^{(2)} \\ 0 & 0 & I \end{pmatrix}.$$

The **Schur complements** are given by $S^{(i)} = A_{BB}^{(i)} - A_{BI}^{(i)} A_{II}^{(i)^{-1}} A_{IB}^{(i)}$, and can each be calculated independently. In general, the matrices $S^{(i)}$ are dense. We now partition the vector of unknown coefficients in the same way as the matrix, $u = (u_I^{(1)}\, u_I^{(2)}\, u_B)^T$. The linear system can then be written as

$$\begin{pmatrix} A_{II}^{(1)} & 0 & 0 \\ 0 & A_{II}^{(2)} & 0 \\ A_{BI}^{(1)} & A_{BI}^{(2)} & I \end{pmatrix} \begin{pmatrix} I & 0 & A_{II}^{(1)^{-1}} A_{IB}^{(1)} \\ 0 & I & A_{II}^{(2)^{-1}} A_{IB}^{(2)} \\ 0 & 0 & S^{(1)} + S^{(2)} \end{pmatrix} \begin{pmatrix} u_I^{(1)} \\ u_I^{(2)} \\ u_B \end{pmatrix} = \begin{pmatrix} f_I^{(1)} \\ f_I^{(2)} \\ f_B \end{pmatrix}.$$

Now perform a (block) forward solve to obtain

$$\begin{pmatrix} I & 0 & A_{II}^{(1)^{-1}} A_{IB}^{(1)} \\ 0 & I & A_{II}^{(2)^{-1}} A_{IB}^{(2)} \\ 0 & 0 & S^{(1)} + S^{(2)} \end{pmatrix} \begin{pmatrix} u_I^{(1)} \\ u_I^{(2)} \\ u_B \end{pmatrix} = \begin{pmatrix} I & 0 & 0 \\ 0 & I & 0 \\ -A_{BI}^{(1)} & -A_{BI}^{(2)} & I \end{pmatrix} \begin{pmatrix} A_{II}^{(1)^{-1}} & 0 & 0 \\ 0 & A_{II}^{(2)^{-1}} & 0 \\ 0 & 0 & I \end{pmatrix} \begin{pmatrix} f_I^{(1)} \\ f_I^{(2)} \\ f_B \end{pmatrix}.$$

We next must solve the **reduced** Schur complement problem,

$$(S^{(1)} + S^{(2)}) u_B = f_B - A_{BI}^{(1)} A_{II}^{(1)^{-1}} f_I^{(1)} - A_{BI}^{(2)} A_{II}^{(2)^{-1}} f_I^{(2)} \equiv g.$$

This procedure can be generalized to the case of N subdomains.

Just as in the previous chapters let R_i be the restriction matrix (consisting of zeros and ones) for the subdomain Ω_i. That is $u^{(i)} = R_i u$ returns the vector of all the coefficients associated with Ω_i. The matrix R_i^T is used to describe the standard assembly process in finite element computations. Since we are interested in the interface variables u_B, we partition R_i into a part acting on the interior variables and a part $\tilde{R}_i$ acting only on the interface variables. The restriction matrix $\tilde{R}_i$ returns the vector of coefficients on the interface of Ω_i. In one dimension this is only two vertices, in two dimensions these coefficients are associated with edges and vertices, while the three dimensional case has interface coefficients on faces, edges, and vertices (see Figure 4.12). Since all substructure methods are concerned with the variables defined on interfaces between the substructures, the tilde symbol on the restriction operators will remind us of this fact. The reader should note that we will use this notation in all of Chapter 4; the interior unknowns are always assumed to be calculated directly (and in parallel) once the interface values on all substructures are known.

Using this, we see how the interface vector u_B can be assembled from the individual substructures as

$$u_B = \sum_{i=1}^{N} \tilde{R}_i^T u_B^{(i)}.$$

The reduced Schur complement problem now takes the general form,

$$\left(\sum_{i=1}^{N} \tilde{R}_i^T S^{(i)} \tilde{R}_i \right) u_B = f_B - \sum_{i=1}^{N} \tilde{R}_i^T A_{BI}^{(i)} A_{II}^{(i)^{-1}} \tilde{R}_i f_I^{(i)} \equiv g. \tag{4.1}$$

The solution of this equation gives us the values along the interior boundary and the problem then splits into independent subproblems. We can therefore backsolve in parallel for the interior unknowns in each substructure,

$$\begin{aligned} u_I^{(i)} &= A_{II}^{(i)^{-1}} f_I^{(i)} - A_{II}^{(i)^{-1}} A_{IB}^{(i)} u_B^{(i)}, \\ &= A_{II}^{(i)^{-1}} (f_I^{(i)} - A_{IB}^{(i)} u_B^{(i)}). \end{aligned}$$

When the Schur complements in (4.1) are explicitly computed we have the direct method called **substructuring**. The engineering community (thinking in terms of elements as the fundamental unit) quickly applied the idea of substructuring in a recursive fashion. As Przemieniecki points out, the Schur complements can be viewed as element stiffness matrices corresponding to complex finite elements where all the interior degrees of freedom have been eliminated by **static condensation**. Equation (4.1) corresponds to a higher level subassembly of a new stiffness matrix. In general, more than two Schur complements will be assembled and the resulting problem can again be divided into substructures, as in Algorithm 4.1.1.

If a structure is modeled with several identical substructures, then the corresponding Schur complement need be computed only once. The repeated Schur complements can be derived from a "reference Schur complement" by a change of basis corresponding to the necessary translation and rotation of the individual substructures just as element stiffness matrices often are computed by using a standard "reference element." This can result in considerable computational savings. Referring to Figure 4.3, but now thinking

Algorithm 4.1.1: Direct Substructuring

- Partition the domain into a hierarchy of nonoverlapping subdomains. Usually the domain is obtained by an assembly of superelements, which are in turn obtained by assembly of smaller superelements.
- Compute the Schur complements of the lowest level superelements.
- Merge the Schur complements.
- Repeat the process until a single Schur complement is obtained.
- Factor the final Schur complement.
- Backsolve in the superelements while descending the hierarchy to obtain the final solution at all nodes.

in terms of substructures, we can construct an alternate, element based "elimination tree" showing the dependencies of the Schur complement computations. This *substructure elimination tree* is shown in Figure 4.4. A more realistic engineering problem is pictured in the same way in Figure 4.5. Note that this decomposition contains identical substructures.

This way of organizing a sparse, direct factorization is very similar to the so-called multifrontal methods that are now widely accepted for general sparse matrix factorizations. Briefly, one may say that the structural analysis community adopted and refined the multilevel substructuring technique in the 1960s, but in a completely element (substructure) oriented way, while the modern multifrontal codes were developed later with the nodes as the fundamental object. Computationally this means that the multifrontal codes always work with competely dense front matrices. The fronts move toward the root of the elimination tree (as in Figure 4.3), merging and growing in size along the way. The multilevel substructure or superelement technique, on the other hand, organizes the elimination according to substructure elimination trees as shown in Figures 4.4 and 4.5. This implies that the Schur complements may contain some zeros. The substructures are usually defined from a partitioning of the physical structure. When more than two substructures are merged by combining the corresponding Schur complements, the resulting stiffness matrix often has a block sparsity structure not explicitly present in the multifrontal computational procedure. An advantage of the substructure approach is that it uses explicit knowledge of the underlying geometry to avoid unnecessary computation of identical substructures.

Still, a multifrontal code and a multilevel substructure code share many features and can be considered close relatives. The computation starts at the leaves of the elimination tree with many smaller tasks. At this stage there is a lot of parallel computation that can take place; typically each processor can work on a sparse matrix corresponding to a first level substructure. When the computation proceeds toward the root of the tree, there are fewer parallel tasks, but each task becomes a large, dense linear algebra problem that can be solved by using a parallel algorithm. Similarly, in the back substitution phase

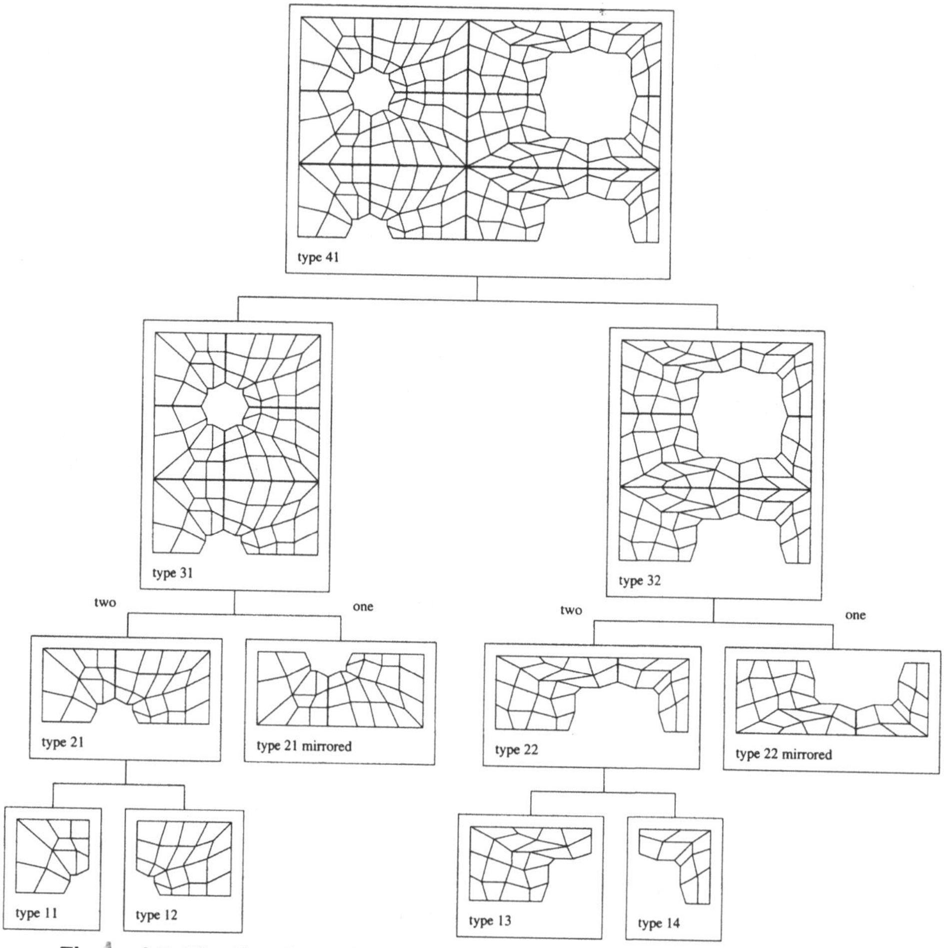

Figure 4.5. The Four Level Substructure Elimination Tree of a Mechanical Part

the process starts out with a large dense triangular solve and quickly proceeds to a stage with many independent substructures, providing each processor a separate triangular back substitution to compute. Note that in this phase the identical substructures must be treated separately, thus giving rise to even more parallel work.

We conclude this section on direct substructuring methods by describing a large scale computation of an offshore oil production platform from the North Sea. Figure 4.6 shows the complete model and Figure 4.7 shows the corresponding substructure elimination tree. This structure has 903,354 unknowns; it consists of 269 first level substructures, of which 100 are different. The elimination tree has eight levels of substructures and there are 39 different substructures above the first level. The triangles in the elimination tree indicate repeated substructures whose Schur complements are already defined elsewhere. In this particular finite element analysis, the linear system of equations had 82 different right hand side vectors. The direct factorization of this problem was initially performed on a Cray X-MP in 1988 and required more than 50 hours of CPU time distributed over many weeks of elapsed analysis time. The problem still requires

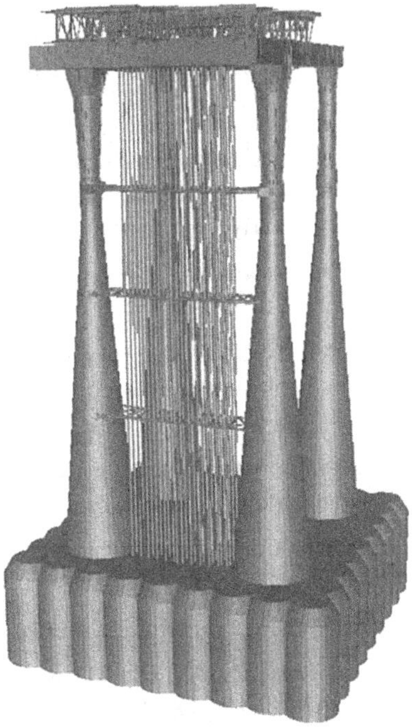

Figure 4.6. Oil Production Platform Solved with Substructuring

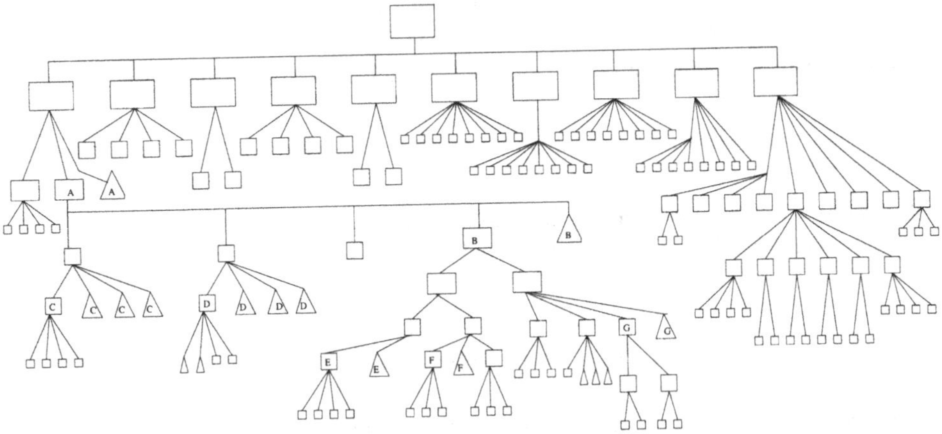

Figure 4.7. Substructure Elimination Tree for Oil Platform

several hours of elapsed computing time and about 20 gigabytes of storage on 1995 supercomputers.

The direct substructuring method and the large example described above should serve as motivation for the iterative methods that are the subject of the rest of this chapter. A thorough understanding of existing methods based on substructuring is important in order to advance hybrid or fully iterative techniques as effective alternatives to the classical methods.

Notes and References for Section 4.1

- Przemieniecki developed the direct substructure method on the basis of assembly of Schur complement matrices in 1961; this work was published in 1963 [Prz63]. In his work the term substructuring and the systematic use of superelements are well established. The use of substructure concepts also appears in early work by Argyris and Kelsey [Arg59], [Arg60].
- See George and Liu [Geo81] for an introduction to sparse matrix techniques for the solution of symmetric linear systems.
- Chapter 9 of Przemieniecki's book from 1968, *Theory of Matrix Structural Analysis* [Prz85], is devoted to substructuring. This book contains more than 350 references to early work in matrix structural analysis.
- For a good reference on multifrontal methods see, for example, Duff, Erisman, and Reid [Duf86], and the more recent survey article by Liu [Liu92].
- Reid [Rei84] describes an implementation of a multifrontal code that also mentions the substructuring concept.
- Bjørstad discusses the parallel implementation of direct substructuring methods in [Bjø87].
- A large commercial finite element code, SESAM [Bel73], uses the substructuring approach. The first release of this code, SESAM'69 [Ber68], [Krå74], was perhaps the first completely recursive implementation of the substructure/superelement solution technique. In the last part of the 1980s a new version of this software was modified to run in parallel, see Hvidsten [Hvi90].
- Hvidsten describes a parallel solution of the large oil platform example using a set of powerful workstations in [Hvi93].
- Elimination trees were first introduced by Schreiber [Sch82]; a comprehensive survey of their use is given in [Liu90].
- References for popular bandwidth and profile minimization algorithms are Crane, Gibbs, and Poole [Cra76] and Gibbs [Gib76].
- Reordering algorithms for sparse factorizations including parallel implementations are discussed in Liu [Liu89] and Lewis, Peyton, and Pothen [Lew89].

4.2 The Two Subdomain Case

The explicit calculation of Schur complements is expensive and requires a large amount of memory since the Schur complements are denser (though of much smaller dimension) than the original stiffness matrices. In iterative substructuring (or Schur complement) methods often the Schur complements need not be explicitly formed.

Recall that the Schur complement is given by

$$S^{(i)} = A_{BB}^{(i)} - A_{BI}^{(i)} A_{II}^{(i)^{-1}} A_{IB}^{(i)}.$$

Note that the action of the Schur complement, $S^{(i)}$, on a vector can be calculated implicitly with three sparse matrix vector multiplies and one matrix solve. In iterative substructuring methods, the linear system

$$Su_B = \left(\sum_{i=1}^{N} \tilde{R}_i^T S^{(i)} \tilde{R}_i \right) u_B = g$$

is solved iteratively, using a preconditioned Krylov subspace method.

Since the convergence rate of iterative methods depends on the condition number of the matrices (see Appendix 1), we digress for a moment to discuss the relationship between the condition number of a symmetric positive definite matrix

$$\begin{pmatrix} A_{II} & A_{IB} \\ A_{BI} & A_{BB} \end{pmatrix}$$

and its Schur complement. Consider the smallest eigenvalue of A expressed by using the Rayleigh quotient of A,

$$\lambda_{\min} = \min_{x \neq 0} \frac{x^T A x}{x^T x}.$$

Now, let us restrict x to the subspace defined by the constraint

$$A_{II} x_I + A_{IB} x_B = 0.$$

In this subspace we can express $x_I = -A_{II}^{-1} A_{IB} x_B$ and compute the Rayleigh quotient

$$x^T A x = x_B^T (A_{BB} - A_{IB}^T A_{II}^{-1} A_{IB}) x_B = x_B^T S x_B.$$

Thus

$$\min_{x_B \neq 0} \frac{x_B^T S x_B}{x_B^T x_B} = \min_{\substack{x_B \neq 0; \\ A_{II}x_I + A_{IB}x_B = 0}} \frac{x^T A x}{x_B^T x_B} \geq \min_{x \neq 0} \frac{x^T A x}{x^T x}.$$

Similarly, the maximum of the Rayleigh quotient for S satisfies a similar property. Therefore the condition number of a Schur complement for any symmetric, positive definite matrix is no larger than that of the original matrix.

For the discretization of PDEs, using the finite element or finite difference method, the conditioning of S is usually much better than that of the original matrix A. For instance, for many second order elliptic PDEs, if h is the mesh size, then the condition number of S grows only like $O(1/h)$, while the condition number of A grows like $O(1/h^2)$. In general, because of the large computational expense of dealing with the Schur complement, the resulting decrease in iteration counts does not, in itself, make it worth simply iterating directly on S rather than on A. Therefore, we need to construct efficient preconditioners for the Schur complement.

In the two subdomain case, we wish to construct preconditioners for $S = S^{(1)} + S^{(2)}$ without explicitly forming either $S^{(1)}$ or $S^{(2)}$. This problem attracted great interest for several years and many approaches, none of which is completely satisfactory, have been proposed. These preconditioners are referred to as **interface solvers** or **interface preconditioners**. We will later use these interface preconditioners in the construction of preconditioners for the case of many subdomains. In this general case, the interface preconditioners are for a single edge (or face in three dimensions) shared by exactly two subdomains.

Notes and References for Section 4.2

- A survey article on Schur complements is [Cot74].
- Mandel [Man90b] discusses the quality of block diagonal and Schur complement preconditioning. He proves that if A is positive definite symmetric of the form

$$\begin{pmatrix} A_{II} & A_{IB} \\ A_{BI} & A_{BB} \end{pmatrix}$$

and one preconditions the system with a matrix

$$\begin{pmatrix} A_{II}^{-1} & 0 \\ 0 & B_S \end{pmatrix}$$

then an even better conditioned problem is obtained by preconditioning the Schur complement with respect to A_{II} by the matrix B_S.

- The fact that the condition number of the Schur complement generally improves from $O(h^{-2})$ to $O(h^{-1})$ for discrete, second order elliptic systems is shown, for example, in [Bjø86]. For some model problems, such results can be shown by a direct calculation, see also Section 4.2.4.
- When A is symmetric positive definite an explicit computation of the Schur complement can be carried out in three steps. First, form the Cholesky factor L of A_{II}. Next, solve the system $LX = A_{IB}$. Finally, compute $S = A_{BB} - X^T X$. A numerically different method is to evaluate the expression $A_{BB} - A_{IB}^T L^{-T} L^{-1} A_{IB}$ from right to left. Experience from structural analysis indicates that the former method is slightly faster, while the latter method can be used when storage is at a premium by evaluating the expression one (block) column at a time. The matrix X, which is much denser than A_{IB}, is not needed in this case.

4.2.1 The Neumann-Dirichlet Algorithm

For a scalar, constant coefficient, second order elliptic PDE using a uniform mesh and with two equal sized subdomains that are mirror images of each other, it follows from symmetry that $S^{(1)} = S^{(2)} = \frac{1}{2}S$. This means that either $S^{(1)}$ or $S^{(2)}$ would make an ideal preconditioner for S. Under more general circumstances, it can still be shown that $S^{(i)}$ can be a good preconditioner for S, though the convergence rate will depend weakly on the relative sizes of the two subdomains and on the PDE.

In the Neumann-Dirichlet algorithm one preconditions S by $S^{(1)^{-1}}$, that is, one solves

$$(S^{(1)} + S^{(2)})u_B = (S^{(1)} + S^{(2)})S^{(1)^{-1}} w_B = (I + S^{(2)} S^{(1)-1}) w_B = g \qquad (4.2)$$

by a Krylov subspace method. The action of $S^{(1)^{-1}}$ can be calculated, without explicitly forming $S^{(1)}$, by using the following observation: $A^{(1)}$ may be factored as

$$A^{(1)} = \begin{pmatrix} I & 0 \\ A_{BI}^{(1)} A_{II}^{(1)^{-1}} & I \end{pmatrix} \begin{pmatrix} A_{II}^{(1)} & 0 \\ 0 & S^{(1)} \end{pmatrix} \begin{pmatrix} I & A_{II}^{(1)^{-1}} A_{IB}^{(1)} \\ 0 & I \end{pmatrix}.$$

Therefore

$$A^{(1)^{-1}} = \begin{pmatrix} I & -A_{II}^{(1)^{-1}} A_{IB}^{(1)} \\ 0 & I \end{pmatrix} \begin{pmatrix} A_{II}^{(1)^{-1}} & 0 \\ 0 & S^{(1)^{-1}} \end{pmatrix} \begin{pmatrix} I & 0 \\ -A_{BI}^{(1)} A_{II}^{(1)^{-1}} & I \end{pmatrix}$$

or

$$A^{(1)^{-1}} = \begin{pmatrix} \cdots & \cdots \\ \cdots & S^{(1)^{-1}} \end{pmatrix},$$

which means that

$$S^{(1)^{-1}} v = \begin{pmatrix} 0 & I \end{pmatrix} A^{(1)^{-1}} \begin{pmatrix} 0 \\ I \end{pmatrix} v. \tag{4.3}$$

Hence we can obtain the action of $S^{(1)^{-1}}$ on a vector at the expense of solving a system involving $A^{(1)}$. This linear system corresponds to solving the discrete problem in Ω_1 with a Neumann boundary condition on the interior interface. The application of $S^{(2)}$ on a vector requires the solution of the linear system $A_{II}^{(2)}$, which corresponds to a discrete problem on Ω_2 with Dirichlet boundary conditions on the artificial boundary, hence the name Neumann-Dirichlet method.

The Neumann-Dirichlet method can be related directly to the original linear system

$$\begin{pmatrix} A_{II}^{(1)} & A_{IB}^{(1)} & 0 \\ A_{BI}^{(1)} & A_{BB}^{(1)} + A_{BB}^{(2)} & A_{BI}^{(2)} \\ 0 & A_{IB}^{(2)} & A_{II}^{(2)} \end{pmatrix} \begin{pmatrix} u_I^{(1)} \\ u_B \\ u_I^{(2)} \end{pmatrix} = \begin{pmatrix} f_I^{(1)} \\ f_B \\ f_I^{(2)} \end{pmatrix}. \tag{4.4}$$

If we first solve two independent subproblems

$$A_{II}^{(i)} v_I^{(i)} = f_I^{(i)} \qquad i = 1, 2,$$

and then set $u_I^{(i)} = v_I^{(i)} + w_I^{(i)}$. The remaining equations for $w_I^{(i)}$ will now have zero in place of $f_I^{(i)}$, that is,

$$\begin{pmatrix} A_{II}^{(1)} & A_{IB}^{(1)} & 0 \\ A_{BI}^{(1)} & A_{BB}^{(1)} + A_{BB}^{(2)} & A_{BI}^{(2)} \\ 0 & A_{IB}^{(2)} & A_{II}^{(2)} \end{pmatrix} \begin{pmatrix} w_I^{(1)} \\ u_B \\ w_I^{(2)} \end{pmatrix} = \begin{pmatrix} 0 \\ f_B \\ 0 \end{pmatrix}.$$

Let us therefore see how the Neumann-Dirichlet preconditioner behaves when applied to a vector y restricted to the internal boundary of our domain. This corresponds to solving the block triangular system

$$\begin{pmatrix} A_{II}^{(1)} & A_{IB}^{(1)} & 0 \\ A_{BI}^{(1)} & A_{BB}^{(1)} & 0 \\ 0 & A_{IB}^{(2)} & A_{II}^{(2)} \end{pmatrix} \begin{pmatrix} x_I^{(1)} \\ x_B \\ x_I^{(2)} \end{pmatrix} = \begin{pmatrix} 0 \\ y_B \\ 0 \end{pmatrix}. \tag{4.5}$$

Block elimination gives

$$x_I^{(i)} = -A_{II}^{(i)^{-1}} A_{IB}^{(i)} x_B \qquad i = 1, 2,$$

with

$$x_B = S^{(1)^{-1}} y_B.$$

We then multiply this vector by the matrix of our original linear system

$$\begin{pmatrix} A_{II}^{(1)} & A_{IB}^{(1)} & 0 \\ A_{BI}^{(1)} & A_{BB}^{(1)} + A_{BB}^{(2)} & A_{BI}^{(2)} \\ 0 & A_{IB}^{(2)} & A_{II}^{(2)} \end{pmatrix} \begin{pmatrix} x_I^{(1)} \\ x_B \\ x_I^{(2)} \end{pmatrix}.$$

The first and last components of this vector are

$$A_{II}^{(i)} x_I^{(i)} + A_{IB}^{(i)} x_B = 0 \qquad i = 1, 2,$$

while the value on the interior boundary becomes

$$\begin{aligned} &-A_{BI}^{(1)} A_{II}^{(1)^{-1}} A_{IB}^{(1)} x_B + A_{BB}^{(1)} x_B + A_{BB}^{(2)} x_B - A_{BI}^{(2)} A_{II}^{(2)^{-1}} A_{IB}^{(2)} x_B \\ &\quad = (S^{(1)} + S^{(2)}) x_B = (S^{(1)} + S^{(2)}) S^{(1)^{-1}} y_B. \end{aligned}$$

Written in terms of our block matrices, we have shown that

$$\begin{pmatrix} A_{II}^{(1)} & A_{IB}^{(1)} & 0 \\ A_{BI}^{(1)} & A_{BB}^{(1)} + A_{BB}^{(2)} & A_{BI}^{(2)} \\ 0 & A_{IB}^{(2)} & A_{II}^{(2)} \end{pmatrix} \begin{pmatrix} A_{II}^{(1)} & A_{IB}^{(1)} & 0 \\ A_{BI}^{(1)} & A_{BB}^{(1)} & 0 \\ 0 & A_{IB}^{(2)} & A_{II}^{(2)} \end{pmatrix}^{-1} \begin{pmatrix} 0 \\ y_B \\ 0 \end{pmatrix} = \begin{pmatrix} 0 \\ (S^{(1)} + S^{(2)}) S^{(1)^{-1}} y_B \\ 0 \end{pmatrix}.$$

The block triangular form of the preconditioner clearly shows how the actual implementation leads to the solution of a Neumann problem on Ω_1 followed by a Dirichlet problem on Ω_2. This method, using only $A_{BB}^{(1)}$, could be viewed as a rather unusual block Gauss-Seidel iteration. The preconditioners, however, are derived from a deeper analysis of the problem than simply an algebraic viewpoint.

We also note that despite the unsymmetric form of the preconditioner, if our original linear system is symmetric then its restriction to the internal boundary results in a symmetric preconditioner that can be accelerated by using the conjugate gradient method.

The Neumann-Dirichlet method as stated in Equation (4.2) and equivalently above leads to a right preconditioning of the matrix. The two methods are described in Algorithm 4.2.1 and Algorithm 4.2.2, respectively. There is a corresponding Dirichlet-Neumann method in which a Dirichlet problem is solved first followed by a Neumann problem.

$$\begin{pmatrix} A_{II}^{(1)} & A_{IB}^{(1)} & 0 \\ A_{BI}^{(1)} & A_{BB}^{(1)} & A_{BI}^{(2)} \\ 0 & 0 & A_{II}^{(2)} \end{pmatrix}^{-1} \begin{pmatrix} A_{II}^{(1)} & A_{IB}^{(1)} & 0 \\ A_{BI}^{(1)} & A_{BB}^{(1)} + A_{BB}^{(2)} & A_{BI}^{(2)} \\ 0 & A_{IB}^{(2)} & A_{II}^{(2)} \end{pmatrix} \begin{pmatrix} 0 \\ y_B \\ 0 \end{pmatrix} = \begin{pmatrix} \cdots \\ S^{(1)^{-1}} (S^{(1)} + S^{(2)}) y_B \\ \cdots \end{pmatrix}.$$

This formulation gives rise to a left preconditioned system.

The preconditioners B in this section have the special property that the matrix A can be written as $A = B^{-1} + C$, where the matrix vector product Cx is considerably less costly than computing Ax. We would therefore like to work directly with the matrix C and the preconditioner B in our algorithms. As an example consider the Neumann-

Dirichlet method expressed in terms of the original matrices (i.e., without explicitly forming the Schur complement). In this case

$$C = A - B^{-1} = \begin{pmatrix} 0 & 0 & 0 \\ 0 & A_{BB}^{(2)} & A_{BI}^{(2)} \\ 0 & 0 & 0 \end{pmatrix}.$$

Clearly using a matrix vector product with this sparse C is more economical than using the matrix A. Unfortunately, this will require some modifications of the standard conjugate gradient method. With standard left preconditioning one computes Ap where p is the direction of the next update. This would now become $B^{-1}p + Cp$, hence we need to carry an auxiliary vector $q = B^{-1}p$. This vector can be updated together with the update of the search direction p. Similarly one must also change the standard algorithm when employing right preconditioning. For the other Krylov subspace methods where there is no need to separate the action of the original matrix and the preconditioner and to preserve the symmetry, one can avoid such modifications by working directly with the matrix $BA = I + BC$ or $AB = I + CB$, respectively.

For two subdomains the Neumann-Dirichlet method has no inherent parallelism. For model problems, with Dirichlet boundary conditions, it can be proved that the number of iterations required for convergence is bounded independently of h, but does depend on the PDE and the relative sizes of the two domains.

There is an interesting relationship between Schur complement methods and the classical alternating Schwarz method; see Chapter 1. Consider our domain in Figure 4.8, but now with a subdomain $\Omega_3 \subset \Omega_2$ (shaded in the figure) near the internal interface Γ between Ω_1 and Ω_2. Let $S^{(3)}$ be the Schur complement of the corresponding matrix $A^{(3)}$ with respect to the boundary variables on the interface between Ω_1 and Ω_2. Let us analyze the classical alternating Schwarz method having Ω_3 as the overlap region. The residuals corresponding to the interior of the subdomains will become zero after the first iteration and will remain zero. We can therefore restrict our attention to the case where the only nonzero residuals are located along the interior boundaries. In the first half step of the Schwarz algorithm we therefore solve a Dirichlet boundary value problem on $\Omega_1 \cup \Omega_3$. The linear system has the form

$$\begin{pmatrix} A_{11} & 0 & A_{1\Gamma} \\ 0 & A_{33} & A_{3\Gamma} \\ A_{\Gamma 1} & A_{\Gamma 3} & A_{\Gamma\Gamma} \end{pmatrix} \begin{pmatrix} y_1 \\ y_3 \\ y_\Gamma \end{pmatrix} = \begin{pmatrix} 0 \\ 0 \\ c_\Gamma \end{pmatrix}.$$

By block Gaussian elimination we obtain a reduced system of the form

$$(S^{(1)} + S^{(3)})y_\Gamma = c_\Gamma. \tag{4.6}$$

After the completion of a full alternating Schwarz iteration, the boundary Γ is the only place where there is a nonzero residual since the second half-step produces a zero residual inside all of Ω_2. This corresponds to a reduction of the original full linear system (for the entire domain $\Omega_1 \cup \Omega_2$) to a Schur complement matrix $S^{(1)} + S^{(2)}$. As always, we can express the error e_Γ on Γ as the solution of

$$(S^{(1)} + S^{(2)})e_\Gamma = r_\Gamma,$$

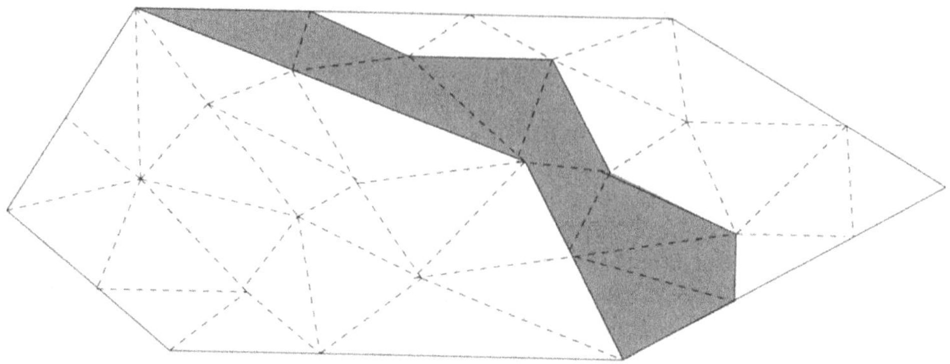

Figure 4.8. The Two Domain Example with Overlap

Algorithm 4.2.1: Neumann-Dirichlet (Interface Only)

- Prepare subdomain Ω_1 to apply the operator $S^{(1)-1}$; note that this requires a solution with Neumann boundary conditions on the interface.
- Prepare subdomain Ω_2 to apply the operator $S^{(2)}$; note that this requires a solution with Dirichlet boundary conditions on the interface.
- Compute the right hand side on the interface by

  $f_B - A_{BI}^{(1)} A_{II}^{(1)^{-1}} f_I^{(1)} - A_{BI}^{(2)} A_{II}^{(2)^{-1}} f_I^{(2)}$.
- Solve the interface problem $(I + S^{(2)} S^{(1)-1}) w_B = g$ with a Krylov subspace method.
- Solve a Dirichlet boundary value problem on each subdomain with w_B on the artificial boundary to obtain the final solution.

where r_Γ is the residual on Γ. But this will be precisely the right hand side c_Γ from (4.6) in our alternating Schwarz iteration. Combining the two expressions immediately shows that the iteration matrix (or equivalently, error propagation operator) is given by

$$I - (S^{(1)} + S^{(3)})^{-1} (S^{(1)} + S^{(2)}).$$

This is equivalent to solving the linear system involving $S^{(1)} + S^{(2)}$ with a preconditioning matrix $B = (S^{(1)} + S^{(3)})^{-1}$. But this implies that the classical overlapping Schwarz iteration on the two domains $\Omega_1 \cup \Omega_3$ and Ω_2 (having Ω_3 as their common overlap) can be viewed as solving the Schur complement system with this special choice of preconditioner. If one could make $S^{(3)} = S^{(1)}$ then this method would also be identical to the Neumann-Dirichlet method. See Algorithms 4.2.1 and 4.2.2. (This construction is possible for some simple model problems by making Ω_3 and its mesh be the mirror image of Ω_1.)

Algorithm 4.2.2: Neumann-Dirichlet (All Variables)

- Prepare subdomain Ω_1 for solution with Neumann boundary conditions on the interface.
- Prepare subdomain Ω_2 for solution with Dirichlet boundary conditions on the interface.
- Compute the solution to the two auxiliary problems, $A_{II}^{(i)} v_i = f_I^{(i)}$.
- Solve the linear system

$$\begin{pmatrix} A_{II}^{(1)} & A_{IB}^{(1)} & 0 \\ A_{BI}^{(1)} & A_{BB}^{(1)} + A_{BB}^{(2)} & A_{BI}^{(2)} \\ 0 & A_{IB}^{(2)} & A_{II}^{(2)} \end{pmatrix} \begin{pmatrix} u_1 \\ u_B \\ u_2 \end{pmatrix} = \begin{pmatrix} 0 \\ f_B \\ 0 \end{pmatrix}$$

with the block triangular right preconditioner

$$\begin{pmatrix} A_{II}^{(1)} & A_{IB}^{(1)} & 0 \\ A_{BI}^{(1)} & A_{BB}^{(1)} & 0 \\ 0 & A_{IB}^{(2)} & A_{II}^{(2)} \end{pmatrix}.$$

- Combine u_i and v_i to obtain the final solution.

Notes and References for Section 4.2.1

- Early discussions of the Neumann-Dirichlet algorithm may be found in Bjørstad and Widlund [Bjø84], [Bjø86].
- The Neumann-Dirichlet method depends on the relative sizes of the subdomains. In [Bjø86] this dependence is explicitly computed for a model T–shaped domain. One can show that the number of iterations depends on the ratio $(m + 1)/(n + 1)$ where m and n are the number of grid points in the two dimensions inside Ω_2. The corresponding result for the J operator (see Section 4.2.4) depends on the aspect ratio of both rectangles.
- Chan and Resasco [Cha87a] proposed the method where $S^{(3)}$ is included in the preconditioner. They showed that this would make the rate of convergence independent of aspect ratios for several model problems.
- Bjørstad and Widlund [Bjø89a] then showed that the Chan-Resasco method was identical to the classical alternating Schwarz iteration. See also Chan and Goovaerts [Cha92a] for a linear algebra proof of the equivalence between the Schur complement and overlapping Schwarz method.

4.2.2 The Neumann-Neumann Method

In the Neumann-Neumann method, the preconditioner is given by $S^{(1)^{-1}} + S^{(2)^{-1}}$. At each iteration, two Dirichlet problems must be solved as well as two Neumann problems. Again the number of iterations required with the Neumann-Neumann algorithm is bounded independently of h, but it does depend on the PDE and the relative sizes of the two domains. In the two subdomain case the Neumann-Neumann method has no advantages over the Neumann-Dirichlet algorithm. It may have more general applicability in the case of many subdomains. The Neumann-Neumann algorithm is discussed in more detail in Sections 4.3.2 and 4.3.3 in the context of several (or many) subdomains.

4.2.3 Hierarchical and Multilevel Methods

In Section 3.1.3, it was observed that a simple change to a hierarchical or multilevel basis can increase the convergence rate greatly at a (relatively) small increase in computational work. These same techniques can be used for the Schur complement on an edge (or face in three dimensions).

Define

$$\text{Schur}\begin{pmatrix} A_{II} & A_{IB} \\ A_{BI} & A_{BB} \end{pmatrix} = A_{BB} - A_{BI}A_{II}^{-1}A_{IB}.$$

A simple calculation shows that

$$\text{Schur}\left(\begin{bmatrix} H_{II}^T & 0 \\ H_{IB}^T & H_{BB}^T \end{bmatrix}\begin{bmatrix} A_{II} & A_{IB} \\ A_{BI} & A_{BB} \end{bmatrix}\begin{bmatrix} H_{II} & H_{IB} \\ 0 & H_{BB} \end{bmatrix}\right) = H_{BB}^T\text{Schur}\begin{pmatrix} A_{II} & A_{IB} \\ A_{BI} & A_{BB} \end{pmatrix} H_{BB}.$$

This implies that if the basis change

$$\begin{pmatrix} H_{II} & H_{IB} \\ 0 & H_{BB} \end{pmatrix}$$

provides a good preconditioner for A then the basis change H_{BB} provides a good preconditioner for Schur(A).

Consider again Figure 4.1 and order the nodes on the interior boundary between the two substructures last. The transformation matrix that represents the change from hierarchical basis to nodal basis will now be upper block triangular with the same structure as our partitioning above. In fact, H_{BB} is the one dimensional basis change from hierarchical to nodal, H_{1d}, along the edge Γ. Therefore, one may precondition the "edge" Schur complement S either symmetrically, as $H_{1d}^T S H_{1d}$, or with left or right preconditioning, $H_{1d}H_{1d}^T S$ or $SH_{1d}H_{1d}^T$. Using the general result that the condition number of the Schur complement of a matrix is bounded by the condition number of that matrix, the simple calculation above and the result for hierarchical basis in two dimensions given in Section 3.1.3, we obtain the fact that for model second order elliptic PDEs,

$$\kappa(H_{1d}H_{1d}^T S) \leq C(1 + \log(H/h))^2.$$

For three dimensions, we obtain the weaker result

$$\kappa(H_{2d}H_{2d}^T S) \leq C\frac{H}{h}(1 + \log(H/h))^2.$$

The same technique may be used to derive preconditioners for the "edge" and "face" Schur complements using the change to multilevel basis preconditioners (BPX) introduced in Chapter 3.1.2. In this case one can obtain preconditioned systems whose condition number is bounded independently of h and H.

Since the preconditioner does not depend on the PDE or the domain on which it is being solved, the convergence for these preconditioners clearly depends on the PDE and on the aspect ratio of the subdomains.

Notes and References for Section 4.2.3

- The algorithm using the change to hierarchical basis along the interface was proposed by Smith and Widlund [Smi90a]. It was inspired by the fact that iterative substructuring algorithms for the p-version finite element method, Babuška, Craig, Mandel, and Pitkäranta [Bab91], which are naturally hierarchically based, do not require interface solves.
- At the same time, Haase, Langer and Meyer [Haa90] independently proposed the same algorithm.
- Inspired by the work of Smith and Widlund, Tong, Chan, and Kuo [Ton91] proposed the use of the multilevel preconditioner (BPX) on the interface.

4.2.4 The J Operator

For the model, constant coefficient problem

$$\begin{aligned} -\Delta u &= f \quad \text{in } \Omega, \\ u &= 0 \quad \text{on } \partial\Omega, \end{aligned}$$

discretized with a uniform mesh on a rectangular subdomain using the standard five point (seven point in three dimensions) finite difference stencil it is possible to derive an explicit formula for the Schur complement $S^{(i)}$. For simplicity, we restrict attention to the case of the union of two square subdomains each with a uniform mesh of $n \times n$ interior mesh points. The mesh width is therefore $h = 1/(n+1)$. Using the natural ordering, both $A_{II}^{(1)}$ and $A_{II}^{(2)}$ are diagonalizable by the tensor product of the one dimensional discrete sine transform,

$$Q_{ij} = \sqrt{\frac{2}{(n+1)}} \sin\left(\frac{ij\pi}{n+1}\right).$$

That is $A_{II}^{(i)} = (Q \otimes Q)(\Lambda \otimes I + I \otimes \Lambda)(Q \otimes Q)$ where the diagonal matrix Λ has components $\Lambda_{ii} = 4\sin^2(\frac{i\pi}{2(n+1)})$. If we define the diagonal matrices

$$\mathcal{A} = I + \tfrac{1}{2}\Lambda - (\Lambda + \tfrac{1}{4}\Lambda^2)^{1/2}$$

and

$$\mathcal{F} = (I - \mathcal{A}^{2n+2})^{-1}(I + \mathcal{A}^{2n+2}),$$

then it can be shown that

$$S^{(i)} = Q\mathcal{F}^{1/2}(\Lambda + \tfrac{1}{4}\Lambda^2)^{1/2}\mathcal{F}^{1/2}Q.$$

This is the explicit eigenvalue decomposition of the Schur complement. We can now easily verify that the condition number of $S^{(i)}$ is indeed of order $O(1/h)$.

By dropping those terms ($\mathcal{F}$) which fall off exponentially fast, and the second order term in Λ, one obtains the simple preconditioner,

$$J_2 = Q\Lambda^{1/2}Q,$$

which can be applied in $O(\frac{1}{h}\log(\frac{1}{h}))$ operations using fast sine transforms. When the mesh is nonuniform, we can simply apply a mapping to a uniform mesh, apply the J operator, and then map back.

In three dimensions for a model domain the interface is a rectangular region with an $n \times m$ mesh, in that case the J operator has the form

$$J_3 = (Q_m \otimes Q_n)(\Lambda_n \otimes I_m + I_n \otimes \Lambda_m)^{1/2}(Q_n \otimes Q_m).$$

When the interface is a general quadrilateral, one would map to a rectangular region, apply the J_3 operator, and then map back. The convergence rate will depend strongly on how distorted the original grid is.

This preconditioner also guarantees a convergence rate that is independent of h, but unfortunately the rate depends on the PDE and on the aspect ratios of the subdomains.

The interface preconditioners in the next two sections try to take the actual PDE into account in the construction of the preconditioner. Neither method has very good convergence properties for small h, but may perform well for moderate h, in situations where the J operator or the hierarchical preconditioner performs poorly, for instance, when the problem has a strong convection term.

Notes and References for Section 4.2.4

- The original use of this technique, in a more abstract setting (the $H^{1/2}(\Gamma)$ space), is due to Dryja [Dry81].
- The explicit derivation and diagonalization of $S^{(i)}$ are due to Bjørstad and Widlund [Bjø84].
- Another derivation was given by Chan [Cha87].
- The J operator technique is also used in Bramble, Pasciak, and Schatz [Bra86].
- Later, Chan and Hou extended the construction to a general constant coefficient five point stencil [Cha91a].
- One can slightly improve on the J operator by also including the next term, that is, use $Q(\Lambda + \Lambda^2/4)^{1/2}Q$. This was suggested by Golub and Mayers [Golu84] and derived by moving the true boundaries that are parallel to Γ to infinity. The connection of this operator to the Neumann-Dirichlet algorithm was observed in Bjørstad and Widlund [Bjø84].
- These methods are closely related to the fictitious domain methods which have been extensively studied in the Soviet Union. See, for instance, Matsokin and Nepomnyaschikh [Mats88], [Mats89] and Nepomnyaschikh [Nep89], [Nep91b].

- An early motivation for the development of domain decomposition methods was the discovery of fast elliptic solvers in the late 1960s–early 1970s for rectangular regions. Domain decomposition ideas were applied to allow these fast solvers to be used on more general regions; see, for example, Concus, Golub, and O'Leary [Con76].

4.2.5 Tangential Components

Recall that $S = A_{BB} - A_{BI}A_{II}^{-1}A_{IB}$. One could propose simply using A_{BB} as a preconditioner. The problem with this approach is that A_{BB} generally is very diagonally dominant while S is not. A_{BB} is diagonally dominant because its diagonal elements include contributions from the components of the operator normal to the edge or face, Γ, but these contributions are not reflected in any off-diagonal elements of A_{BB}. An ad hoc preconditioner for S is to use $A_{BB}^{\top}$, the tangential component of A_{BB}, as a preconditioner for S. The operator $A_{BB}^{\top}$ involves only coupling along the edge or face and not the components associated with the normal direction to the edge or face. It, therefore, is much less diagonally dominant and more likely to reflect the behavior of S.

Numerical experiments have shown that this preconditioner works well in some situations; see Notes and References for Section 4.2.6. Performance of this type of preconditioner will depend strongly on the direction of any convection term. If the convection is along the edge (or face in three dimensions) then S is nearly equal to $A_{BB}^{\top}$ and the tangential preconditioner can be very effective. When the convection is normal to the edge (or face in three dimensions) the preconditioner can perform very badly. It is possible to perform corrections for the normal component that are somewhat effective at improving the convergence when the convection is normal to the face or edge.

4.2.6 Probing

We will often in the construction of preconditioners replace various Schur complements with suitable approximations that are more easily computed. We have already seen that the J operator or a tangential component operator could serve such a purpose. We will denote such an approximation to a Schur complement S by $\hat{S}$ without necessarily specifying which approximation is being used.

It is known for elliptic PDEs that the magnitude of the off-diagonal elements of S decay rapidly away from the diagonal. This suggests using the inverse of the tridiagonal (or pentadiagonal, etc.) part of S as a preconditioner. The tridiagonal part of S is not known; however, it is possible to approximate it by the use of **probing**. We will now demonstrate how one may construct a symmetric tridiagonal approximation to a matrix S. Assume S is approximately equal to $\hat{S}$, where

$$\hat{S} = \begin{pmatrix} a_1 & b_1 & & \\ b_1 & a_2 & b_2 & \\ & b_2 & a_3 & b_3 \\ & & \cdots & \end{pmatrix}.$$

Let $u^1 = (1, \ldots, 1)^T$, and $u^2 = (-1, 1, -1, 1, \ldots)^T$; then calculate $v^1 = Su^1$ and

$v^2 = Su^2$. If, in fact, S equals $\hat{S}$, then from the ith row of $\hat{S}$ we obtain

$$b_{i-1} + a_i + b_i = v_i^1$$

and

$$(-1)^i(-b_{i-1} + a_i - b_i) = v_i^2,$$

for $i = 1, \ldots, n$, and where $b_0 = b_n = 0$. From these equations one easily obtains

$$a_i = \tfrac{1}{2}(v_i^1 + (-1)^i v_i^2) \qquad i = 1, \ldots, n,$$

$$b_i = \tfrac{1}{2}(v_i^1 - (-1)^i v_i^2) - b_{i-1} \qquad i = 1, \ldots, n-1.$$

This technique can be generalized to nonsymmetric problems and also to wider band approximations. For large h this approach can yield an effective preconditioner. The important effect is that one obtains a lumping of the matrix contributions onto the tridiagonal part. The overall condition number can therefore be qualitatively better than what one would obtain by actually using the tridiagonal part of the Schur complement.

Probing can also be used to find other approximations to S. Assuming that S is well approximated by a Toeplitz matrix, one can determine the Toeplitz structure from the first row of S, that is, after a multiplication by the first unit vector. Experience has shown that it is often better to base such an approximation on the values from the middle column of S. This reduces the effect of boundaries and produces a banded Toeplitz matrix (with bandwidth $n/2$). The neglected values far away from the diagonal can most often be set to zero without affecting the performance of the preconditioner. As can be seen below, the resulting preconditioner is often very effective. In order to give an idea of the relative performance of several of the interface preconditioners we list the condition numbers for a model problem below.

Computational Results 4.2.1: Two Subdomain Interface Solvers

Purpose: Compare a variety of interface preconditioners.

PDE: The Poisson equation,

$$-\Delta \boldsymbol{u} = \boldsymbol{f} \quad \text{in } \Omega,$$
$$\boldsymbol{u} = 0 \quad \text{on } \partial\Omega.$$

Domain: Ω is the union of two squares that share one common edge.

Discretization: The PDE was discretized using centered finite differences on a uniform grid. The grid is obtained by uniformly refining the initial grid that had a single node on the artificial boundary. So after one level of refinement there are three nodes along the artificial boundary.

Calculations: The condition numbers were calculated for a variety of the interface preconditioners discussed.

Refinement levels	1	2	5	6	7	k
No. of unknowns on Γ	3	7	63	127	255	$2^{k+1}-1$
Multilevel method (BPX)	1.61	1.95	2.31	2.37	2.41	< 3
The J operator	1.83	2.25	2.44	2.45	2.45	< 3
Hierarchical basis	1.68	2.66	6.75	8.52	10.50	$\approx 0.2 \cdot k^2$
Toeplitz	1.03	1.11	1.45	1.58	1.70	$\approx 0.24 \cdot k$
Tridiagonal probing	1.04	1.21	3.37	4.92	7.16	$\approx 0.62 \cdot 2^{k/2}$
Tridiagonal part	1.04	1.34	8.32	16.58	33.13	$\approx 0.26 \cdot 2^k$
No preconditioning	3.05	6.88	57.37	114.79	230.49	$\approx 1.8 \cdot 2^k$

Discussion: The two first methods have a condition number that increases slightly but, as we have seen, is bounded independently of the discretization. In the last column of the table we list the approximate behavior as a function of the refinement level k. The Toeplitz preconditioner uses the middle column of the Schur complement to define a Toeplitz matrix. This preconditioner is inexpensive since we can use fast Toeplitz solvers to apply the preconditioner. The tridiagonal probing preconditioner is described above. We observe a growth in the condition number that is in agreement with theoretical estimates; see the Notes and References at the end of this section. We see that it does a significantly better job than the tridiagonal part of the matrix. Observe that while the hierarchical basis gives a growth proportional to k^2 the tridiagonal preconditioning produces exponential growth just as in the case with no preconditioning, the only difference is that the constant term is about a factor 7 smaller.

Notes and References for Section 4.2.6

- Numerical experiments for the ad hoc tangential components preconditioner may be found in [Cha90b], [Cai92a], [Gro91], and [Key90b].
- The probing technique was initially developed for estimating sparse Jacobian matrices by Curtis, Powell and Reid [Cur74]. Similar techniques for estimating sparse Hessian matrices are described by Powell and Toint [Pow79].
- In our context probing was used by Chan and Keyes [Cha90b] and Chan and Mathew [Cha92b]. Chan and Mathew proved for a model example that the condition number behaves like $O(h^{-1/2})$.
- The construction of Toeplitz preconditioners via probing was used by Bjørstad in work related to [Bjø80] on the biharmonic equation. The Toeplitz approximation can often be used as an inexpensive preconditioner since there are fast Toeplitz solvers available. The paper by Bunch [Bun85] is a good starting point to the literature on Toeplitz solvers. The intended use as a preconditioner should be kept in mind when considering the stability properties of the algorithms.
- Greenbaum and Rodrigue [Gre89] considered the general problem of constructing optimal

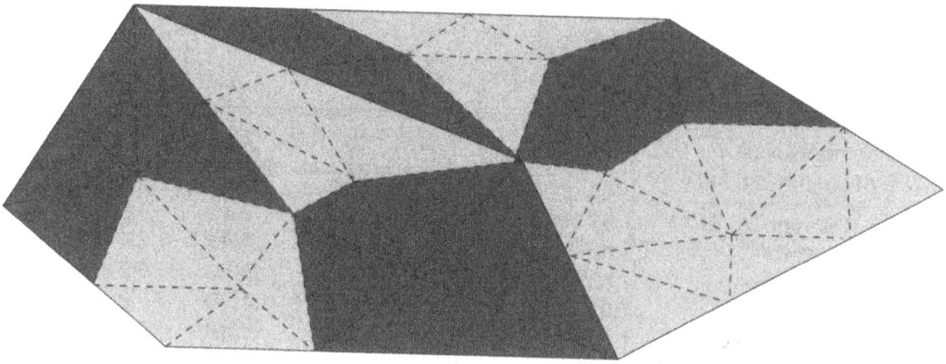

Figure 4.9. Red–Black Coloring

preconditioners with a given sparsity pattern. They show how to attack this problem with general optimization software. On the basis of their findings, they conjecture that the optimal tridiagonal preconditioner to a Schur complement (from a scalar, second order elliptic problem) results in an overall condition number proportional with $h^{-1/2}$. This is in good agreement with the probing numbers in Computational Results 4.2.1 and the theoretical results in [Cha92b].

4.3 Many Subdomains

In this section we present several iterative substructuring methods for use with more than two subdomains. As with the overlapping Schwarz methods it is possible to include a coarse grid problem to enhance the convergence rate greatly. As above, $S^{(i)}$ denotes the Schur complement for subdomain Ω_i. It is important to realize that $S^{(i)}$ is generally dense and contains couplings between the various edges (faces in three dimensions) of the substructures.

4.3.1 One Level Neumann-Dirichlet Methods

When one can make a red-black coloring of the subdomains (see Figure 4.9), it is possible to generalize the Neumann-Dirichlet algorithm to the case of many subdomains. One step of the algorithm requires the solution of Neumann boundary value problems on all of the red subdomains and the solution of Dirichlet boundary value problems on all of the black subdomains. To explore this we need to introduce some additional notations. Let $\tilde{R}_i$ be the restriction operator which maps from the vector of coefficient unknowns on the artificial boundary, u_B, to only those associated with the boundary of Ω_i. We use a superscript to denote the color of subdomain Ω_i. Then the Schur complement can be written as

$$S = \sum_{\Omega_i} \tilde{R}_i^T S_i \tilde{R}_i. \tag{4.7}$$

If

$$S = \sum_{\Omega_i^R} \tilde{R}_i^{(R)^T} S_i^{(R)} \tilde{R}_i^{(R)} + \sum_{\Omega_i^B} \tilde{R}_i^{(B)^T} S_i^{(B)} \tilde{R}_i^{(B)},$$

where we separate the subdomains according to color, then the preconditioner may be written as

$$B = \sum_{\Omega_i^R} \tilde{R}_i^{(R)^T} S_i^{(R)^{-1}} \tilde{R}_i^{(R)}$$

and

$$SB = \sum_{\Omega_i^R} \tilde{R}_i^{(R)^T} \tilde{R}_i^{(R)} + \left(\sum_{\Omega_i^B} \tilde{R}_i^{(B)^T} S_i^{(B)} \tilde{R}_i^{(B)} \right) \left(\sum_{\Omega_i^R} \tilde{R}_i^{(R)^T} S_i^{(R)^{-1}} \tilde{R}_i^{(R)} \right).$$

We note that the first term in the expression for SB is an identity matrix in the case when there are no crosspoints shared by subdomains having the same color, for example, if the domain is sliced into strips of alternating colors.

Notes and References for Section 4.3.1

- In [Dry88] Dryja demonstrated how a "global coarse problem" could be combined with the above method to generate a method whose convergence rate was independent of the number of subdomains.
- In [Bjø88] Bjørstad and Hvidsten performed numerical calculations using the Neumann-Dirichlet algorithm for the finite element solution of the equations of linear elasticity.

4.3.2 One Level Neumann-Neumann Methods

The Neumann-Neumann method has a natural extension to many subdomains. Let D be a diagonal scaling matrix, where for each element of u_B, D_{jj}^{-1} is the number of subdomains which share the node j. Then the Neumann-Neumann preconditioner is given by

$$B = D \left(\sum_{\Omega_i} \tilde{R}_i^T S^{(i)^{-1}} \tilde{R}_i \right) D.$$

More generally, one may use

$$B = \sum_{\Omega_i} D_i \tilde{R}_i^T S^{(i)^{-1}} \tilde{R}_i D_i.$$

Better convergence seems to be obtained when $\sum_{\Omega_i} D_i = I$. Again, it is important to remember that the action of $S^{(i)^{-1}}$ is calculated without explicitly forming $S^{(i)}$, by using (4.3).

If Ω_i is an interior subdomain, then $S^{(i)}$ is a singular matrix. We must therefore apply a pseudoinverse. The convergence of the overall scheme depends strongly on the accuracy of the pseudoinverse used. The inverse which is used for the analysis of the methods is sometimes the inverse of the Schur complement of the matrix $A_i + \alpha I$. One may also use the Moore-Penrose pseudo-inverse, though that is rather expensive computationally. Another approach that may be used when the null vectors of $S^{(i)}$ are known is to solve the linear system in the complement of the null space. For many iterative methods this is relatively simple; it simply means removing any part of each iterate that lies in the null space. For direct methods, a nonsingular matrix can be constructed whose solutions lie in the complement of the null space. One example in which the null space is known is the Poisson equation; in practice, the pressure equation of fluid dynamics is a Poisson problem.

The convergence of the Neumann-Neumann algorithm for a small number of subdomains is quite good. The number of iterations required to achieve a fixed tolerance grows like $(1 + \log(H/h)/H$. However, the convergence rate decays rapidly for large numbers of subdomains, as a result of the $1/H$ term. The balancing algorithm, introduced below in Section 4.3.3, includes a coarse grid problem to reduce this dependence on the number of subdomains significantly.

Notes and References for Section 4.3.2

- The origins of the Neumann-Neumann method can be traced at least as far back as the work of Dinh and Périaux [Dih84]. At that time, they were essentially solving the dual problem with no preconditioning.
- Later studies were continued by Bourgat, Glowinski, Le Tallec, and Vidrascu [Bou89], Le Tallec, De Roeck, and Vidrascu [Le T91], and De Roeck and Le Tallec [De R91]. See also Glowinski and Wheeler [Glo88b] where Schur complement methods are used on mixed finite element problems.
- Additional analysis has been performed by Dryja and Widlund [Dry90], [Dry95].
- More recently Kuznetsov, Manninen and Vassilevski have performed a systematic set of numerical studies [Kuz93].
- It is also possible to consider generalizations of the Neumann-Dirichlet and Neumann-Neumann algorithms by solving subdomain problems with mixed boundary conditions.

4.3.3 Balancing Neumann-Neumann Methods

By introducing a two level structure for the Neumann-Neumann method, we can obtain algorithms with potentially good convergence and little need for geometric information. The balancing Neumann-Neumann method is simply the Neumann-Neumann preconditioner with the addition of a particularly simple coarse grid correction. This correction is constructed by using a piecewise constant coarse grid space and its corresponding operator A_0 defined below.

We first introduce the algorithm as a three step method,

$$\begin{aligned} u^{n+1/3} &\leftarrow u^n + \tilde{R}_0^T A_0^{-1} \tilde{R}_0 (f - Su^n), \\ u^{n+2/3} &\leftarrow u^{n+1/3} + D \sum_{\Omega_i} \tilde{R}_i^T S^{(i)^{-1}} \tilde{R}_i D(f - Su^{n+1/3}), \\ u^{n+1} &\leftarrow u^{n+2/3} + \tilde{R}_0^T A_0^{-1} \tilde{R}_0 (f - Su^{n+2/3}). \end{aligned} \tag{4.8}$$

The preconditioner may be written as

$$B = (I - \tilde{R}_0^T A_0^{-1} \tilde{R}_0 S) D \left(\sum_{\Omega_i} \tilde{R}_i^T S^{(i)^{-1}} \tilde{R}_i \right) D (I - S \tilde{R}_0^T A_0^{-1} \tilde{R}_0) + \tilde{R}_0^T A_0^{-1} \tilde{R}_0.$$

Algorithm 4.3.1: Balancing Neumann-Neumann

- Prepare all subdomains to apply the Schur complement; note that this involves solving a Dirichlet boundary value problem on the artificial boundaries.
- Prepare all subdomains to apply the inverse of the Schur complement; note that this involves solving a Neumann boundary value problem on the artificial boundaries.
- Prepare the piecewise constant restriction and interpolation operators.
- Form $A_0 = \tilde{R}_0 S \tilde{R}_0^T$.
- Form the right hand side for the interface Schur complement problem.
- Solve the Schur complement problem with a preconditioned Krylov subspace method with the preconditioner given by applying (4.8) with zero initial guess.
- Backsolve to obtain the solution on the subdomain interiors.

For nonsymmetric problems the first substep may be dropped (it is needed only to make the operation symmetric).

For scalar, second order elliptic PDEs, the coarse problem, A_0, involves one unknown per subdomain. The restriction operator $\tilde{R}_0$ simply returns for each subdomain the weighted sum of the values of all the nodes on the boundary of that subdomain. The weights are determined by the inverse of the number of subdomains each node is contained in. The matrix A_0 is given by $A_0 = \tilde{R}_0 S \tilde{R}_0^T$.

A nice feature of the balancing approach is that it allows the use of completely unstructured subdomains. The subdomains need not form a coarse triangulation, and, in addition, knowledge of what nodes are on faces, edges, or vertices is not needed to implement the algorithm. It is possible, for certain model problems, to prove that for this algorithm the condition number of the preconditioned problem grows like $O((1 + \log(H/h))^2)$ in both two and three dimensions. In addition, with a slight modification to the scaling in the restriction operator $\tilde{R}_0$ the convergence rate may be bounded independently of the jumps in the coefficients of the PDE between subdomains.

This method is called the balancing algorithm (see Algorithm 4.3.1) because the calculation of $u^{n+1/3}$ and u^{n+1} balances the average of u on each subdomain. This coarse grid correction is actually very similar to the one we will introduce in Section 4.3.5 for the so-called wire basket method. The basic idea is to remove the constant part of the error from each subdomain at each iteration; recall also Section 2.1.

By using piecewise constant functions in the coarse grid space it is possible to have substructures of very general shape. This introduces one of the strengths of the balancing Neumann-Neumann method over the overlapping Schwarz methods. For elliptic PDEs using a piecewise constant coarse grid space is not adequate with the overlapping Schwarz methods. Mathematically, this is because the energy of piecewise constant interpolations in $H^{1/2}(\Gamma)$ is much better behaved than in $H^1(\Omega)$; see Section 5.3.3.

Notes and References for Section 4.3.3

- Combining some of his earlier work [Man90a], [Man90c] (which we will explore later) with the Neumann-Neumann algorithm, Mandel [Man93] and Mandel and Brezina [Man92] essentially introduced the algorithm given above. See also Mandel [Man94].
- Cowsar, Mandel, and Wheeler later modified the algorithm for mixed finite elements [Cow92], [Cow93].
- The algorithm has properties similar to those of the algorithm due to Smith [Smi91], which will be discussed below; cf. also Bramble, Pasciak, and Schatz [Bra89].
- Perhaps the most complete categorization of two level Neumann-Neumann algorithms to date is given in Dryja and Widlund [Dry95].
- Farhat and Roux discuss several domain decomposition algorithms in the monograph [Far94]. They discuss both direct and iterative methods applied to structural mechanics problems. Their method, finite element tearing and interconnecting (FETI), can be considered a dual to the methods that precondition an interface Schur complement problem in the displacement variables. The preconditioning and treatment of the appropriate null space (corresponding to rigid body motions) are similar to the balancing Neumann-Neumann technique. The monograph discusses several large scale, realistic structural mechanics computations using parallel computers.
- Le Tallec discusses domain decomposition in computational mechanics in the monograph [Le T94]. He discusses the mathematical theory of domain decomposition methods with considerable attention to the Neumann-Neumann class of algorithms.

4.3.4 Iterative Substructuring in Two Dimensions

The algorithm to be discussed in this section, the classical iterative substructuring method, does require explicit knowledge of the geometry of the subdomains, unlike the balancing Neumann-Neumann method. The use of many subdomains with Schur complement methods introduces **vertex points.** These points, which are also called cross points, are isolated (corner) points on the interface that belong to the boundary of more subdomains than neighboring interface points. That is, in both two and three dimensions we will refer to the vertices (corners) of the subdomains as vertex points and the edges of the subdomains as **edges**. In addition in three dimensions we will refer to the edges and vertices as the **wire basket**. Figure 4.12 gives an illustration of these concepts. For Schur complement methods it is possible to use the vertex points or the full wire basket in the construction of two level algorithms. We consider the two dimensional case here and proceed to three dimensions in the next section. In the next two sections we restrict attention to scalar elliptic PDEs and piecewise linear finite elements.

We begin by reordering the unknowns on the interface Γ, listing first those that lie on each individual interface (edge), E_{ij}, followed by those that lie on vertex points; see Figure 4.10. The Schur complement, S, can then be written as

$$\begin{pmatrix} S_{EE} & S_{EV} \\ S_{VE} & S_{VV} \end{pmatrix}.$$

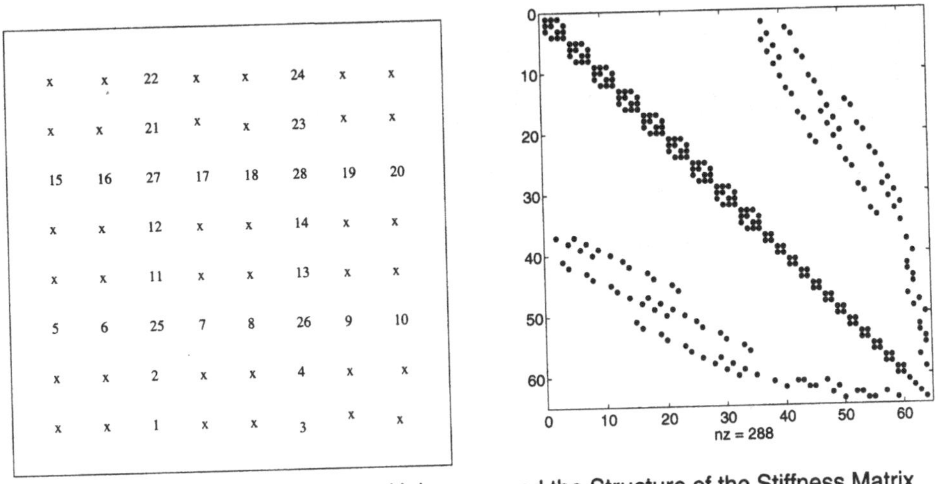

Figure 4.10. Ordering of Interface Unknowns and the Structure of the Stiffness Matrix

In fact, S_{VV} is a diagonal matrix and S_{EE} has large blocks along the diagonal (but is not itself block diagonal), each associated with a different edge E_{ij} (see the left side of Figure 4.11). A natural and simple preconditioner can be constructed for S by simply dropping all couplings between different edges and between edges and vertex points. The resulting matrix structure is shown on the right in Figure 4.11. Then we can replace each block associated with an edge with one of the interface preconditioners, for instance, the J operator, introduced in Section 4.2.4. We recall the notation $\hat{S}$ is used to denote that S has been replaced by an interface preconditioner. Our preconditioner now has the form

$$B = \begin{pmatrix} \hat{S}_{EE}^{-1} & 0 \\ 0 & S_{VV}^{-1} \end{pmatrix}.$$

This is a poor preconditioner, since it is only a one level solver, because S_{VV} is a diagonal matrix which does not couple the various subdomains. We would like to modify the preconditioner so that the vertex points are coupled to the edges and to each other and provide a two level structure. We begin by recalling the coarse grid restriction operator, R, introduced in Section 2.2. Since we are only interested in the interface variables, the R in this chapter is modified so that we only keep the part that acts on the interface unknowns and not on the unknowns associated with the interiors of the subdomains. We denote this modified restriction matrix $\tilde{R}$. (We made a similar change of notation in Section 4.1.) Recall that the transpose, $\tilde{R}^T$, defines an interpolation from the coarse grid to only the fine grid points on the interfaces between substructures. Note that $\tilde{R}$ can be written as $\tilde{R} = (\tilde{R}_E \;\; I)$, using the ordering of the unknowns just introduced. In fact, for linear coarse grid elements the interpolation operator, R^T, acting on a coarse vector of grid points, x_V, can be expressed as

$$R^T x_V = \begin{pmatrix} -A_{II}^{-1} A_{IB} \\ I \end{pmatrix} \tilde{R}_E^T x_V. \tag{4.9}$$

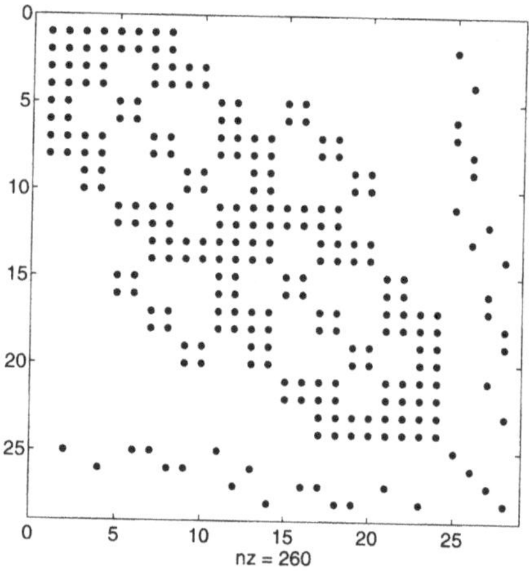

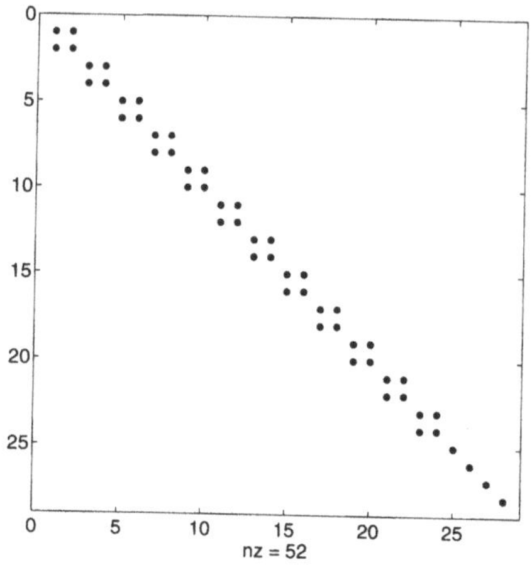

Figure 4.11. Matrix Representation of the Schur Complement and a Simple Block Diagonal Preconditioner

This says that the coarse vertex point values are first interpolated onto the edges of the coarse grid element (substructure), then extended linearly to the interior fine grid points. The matrix

$$\begin{pmatrix} I & \tilde{R}_E^T \\ 0 & I \end{pmatrix}$$

provides a basis change from a **partial hierarchical basis** to the usual nodal basis restricted to the interface. It simply linearly interpolates the values at the vertex points onto the edge nodes.

The Schur complement can be written as

$$\begin{pmatrix} I & 0 \\ -\tilde{R}_E & I \end{pmatrix} \begin{pmatrix} I & 0 \\ \tilde{R}_E & I \end{pmatrix} \begin{pmatrix} S_{EE} & S_{EV} \\ S_{VE} & S_{VV} \end{pmatrix} \begin{pmatrix} I & \tilde{R}_E^T \\ 0 & I \end{pmatrix} \begin{pmatrix} I & -\tilde{R}_E^T \\ 0 & I \end{pmatrix}$$

or

$$\begin{pmatrix} I & 0 \\ -\tilde{R}_E & I \end{pmatrix} \begin{pmatrix} S_{EE} & \tilde{S}_{EV} \\ \tilde{S}_{VE} & \tilde{S}_{VV} \end{pmatrix} \begin{pmatrix} I & -\tilde{R}_E^T \\ 0 & I \end{pmatrix},$$

where $\tilde{S}_{EV}$ denotes that the corresponding matrix block has been modified. Drop the off-diagonal terms which represent the couplings between different edges (changing S_{EE}) and between the edges and the vertex points to obtain

$$\begin{pmatrix} I & 0 \\ -\tilde{R}_E & I \end{pmatrix} \begin{pmatrix} \tilde{S}_{EE} & 0 \\ 0 & \tilde{S}_{VV} \end{pmatrix} \begin{pmatrix} I & -\tilde{R}_E^T \\ 0 & I \end{pmatrix}.$$

The matrix $\tilde{S}_{VV}$ is no longer diagonal, but is a sparse matrix which couples each vertex point (the endpoints of the edges) to the neighboring vertex points.

Since we do not wish to explicitly form S we replace the diagonal blocks and obtain

$$B^{-1} = \begin{pmatrix} I & 0 \\ -\tilde{R}_E & I \end{pmatrix} \begin{pmatrix} \hat{S}_{EE} & 0 \\ 0 & A_V \end{pmatrix} \begin{pmatrix} I & -\tilde{R}_E^T \\ 0 & I \end{pmatrix},$$

which leads to the final preconditioner

$$B = \begin{pmatrix} I & \tilde{R}_E^T \\ 0 & I \end{pmatrix} \begin{pmatrix} \hat{S}_{EE}^{-1} & 0 \\ 0 & A_V^{-1} \end{pmatrix} \begin{pmatrix} I & 0 \\ \tilde{R}_E & I \end{pmatrix}.$$

Here A_V is the same coarse grid operator as A_C introduced in Section 2.2; see also the definition of the Galerkin coarse grid operator A_C in Section 2.4. We use the notation A_V here to remind the reader that this coarse grid matrix is defined on the vertex points and that it replaced the matrix $\tilde{S}_{VV}$. In general, the matrix A_V is spectrally equivalent to $\tilde{S}_{VV}$. For piecewise linear finite elements it follows from (4.9) that $A_V = \tilde{S}_{VV}$. To see this one just computes the Galerkin coarse grid operator $A_V = RAR^T$, observing that $A_V = R_E S_{VV} R_E^T = \tilde{S}_{VV}$.

The interface preconditioner, $\hat{S}_{EE}$, approximates $\tilde{S}_{EE}$ and is a block diagonal matrix with one block for each edge, each block of which is an interface preconditioner from Section 4.2, for instance, the J operator or an operator obtained from probing. Again, these matrices are not formed; instead we need merely to be able to apply the action of the operators onto a vector.

For nonsymmetric problems, where preserving the symmetry of the preconditioner, B, is not important, one can drop either the first or the third factor from the preconditioner.

The preconditioner can be written in a form that makes it look more like the preconditioners introduced in Chapters 1 and 2. Let $\tilde{R}_{ij}$ be the restriction matrix (having zero and one as entries) that returns the values associated with the edge E_{ij}. Just as in Section 4.1, we can use this matrix to express the assembly of $\hat{S}_{EE}$ from individual contributions. (The reader should keep in mind that all restriction matrices in this chapter are defined on the interface variables only, that restriction matrices with lowercase subscripts are zero–one matrices that help define the assembly of substructures, and that $\tilde{R}$ and restriction matrices subscripted by uppercase letters represent the transpose of an interpolation operator from a coarse mesh, defined by substructures, onto the fine mesh grid points on the interface.) If B_{ij} represents the interface preconditioner associated with the edge, E_{ij}, then

$$\hat{S}_{EE}^{-1} = \begin{pmatrix} I_E \\ 0 \end{pmatrix} \left(\sum_{E_{ij}} \tilde{R}_{ij}^T B_{ij} \tilde{R}_{ij} \right) \begin{pmatrix} I_E \\ 0 \end{pmatrix}^T$$

and

$$B = \tilde{R}^T A_V^{-1} \tilde{R} + \sum_{E_{ij}} \tilde{R}_{ij}^T B_{ij} \tilde{R}_{ij}.$$

We refer to B as a **direct sum preconditioner** because the sum of the dimensions of A_V and B_{ij} equals the dimension of S.

Though this preconditioner was derived by using block matrix notation, it can, in fact, be interpreted as a two level overlapping Schwarz method with a very particular choice

Algorithm 4.3.2: Classical Iterative Substructuring

- Prepare each subdomain to apply the operator $S^{(i)}$.
- Construct the coarse grid matrix, A_V.
- For each edge, construct an edge preconditioner B_{ij}, using, for instance, the J operator.
- Solve the Schur complement problem with preconditioner $\tilde{R}^T A_V^{-1} \tilde{R} + \sum_{E_{ij}} \tilde{R}_{ij}^T B_{ij} \tilde{R}_{ij}$.
- Backsolve to obtain the subdomain interior unknowns.

for the overlapping subdomains. Each subdomain is the union of two adjacent coarse grid elements that share a common edge (face in three dimensions). These individual problems are then solved by combining two subdomain solves and the corresponding interface solve. This is demonstrated in Section 4.6.

We now describe the important **null space property** of the iterative substructuring algorithm (see Algorithm 4.3.2). This property is important since it allows us to derive global estimates from the local properties of Schur complements and their interface preconditioners associated with the boundary of individual substructures. We express the assembly of S just as in (4.7),

$$S = \sum_{\Omega_i} \tilde{R}_i^T S^{(i)} \tilde{R}_i,$$

so the inverse of the preconditioner, B^{-1}, may be written as

$$B^{-1} = \sum_{\Omega_i} \tilde{R}_i^T B^{(i)^{-1}} \tilde{R}_i,$$

where

$$B^{(i)^{-1}} = \begin{pmatrix} I & 0 \\ -\tilde{R}_{E^i} & I \end{pmatrix} \begin{pmatrix} \hat{S}_{EE}^{(i)} & 0 \\ 0 & A_{VV}^{(i)} \end{pmatrix} \begin{pmatrix} I & -\tilde{R}_{E^i}^T \\ 0 & I \end{pmatrix}.$$

For all subdomains Ω_i, the null spaces of $S^{(i)}$ and $B^{(i)}$ are identical. (Recall that in this section we consider scalar PDEs only. The null space will correspond to constant functions and linear interpolation reproduces constants exactly.) If there exist constants c_i and C_i such that

$$c_i u^T \tilde{R}_i^T B^{(i)^{-1}} \tilde{R}_i u \le u^T \tilde{R}_i^T S^{(i)} \tilde{R}_i u \le C_i u^T \tilde{R}_i^T B^{(i)^{-1}} \tilde{R}_i u,$$

then, by summing over the subdomains,

$$(\min_i c_i) u^T B^{-1} u \le u^T S u \le (\max_i C_i) u^T B^{-1} u.$$

It follows that the convergence rate when using B as a preconditioner for S will not depend explicitly on the number of subdomains, only on c_i and C_i, which may be calculated locally on each individual substructure.

Consider the case of jump coefficients, where

$$a(\boldsymbol{u}, \boldsymbol{v}) = \sum_i \int_{\Omega_i} \rho_i(x)(\nabla \boldsymbol{u}, \nabla \boldsymbol{v})\, dx$$

and $\rho_i(x)$ is smooth inside the subdomains but may have large jumps between subdomains. Define $B^{(i)} = \bar{\rho}_i \tilde{B}^{(i)}$ where $\tilde{B}^{(i)}$ is the preconditioner associated with Ω_i for the Laplacian, and $\bar{\rho}_i$ is the average of $\rho_i(x)$ on Ω_i; then there exist constants c_i and C_i independent of the jumps in $\rho_i(x)$ so that

$$c_i u^T \tilde{R}_i^T B^{(i)^{-1}} \tilde{R}_i u \le u^T \tilde{R}_i^T S^{(i)} \tilde{R}_i u \le C_i u^T \tilde{R}_i^T B^{(i)^{-1}} \tilde{R}_i u.$$

Thus the convergence rate for this scheme is also bounded independently of the jumps in $\rho_i(x)$.

In two dimensions, the theoretical and numerically observed condition numbers for these methods grow as $O((1 + \log(H/h))^2)$, for certain model problems; see Sections 4.6 and 5.3.2. Hence the number of iterations of a Krylov subspace method to achieve a fixed tolerance grows like $(1 + \log(H/h))$. These methods are not optimal, but since they behave well for problems with large jumps in the PDE's coefficients between subdomains, they have been studied extensively. The convergence of these algorithms in three dimensions is discussed in the next section.

In order to demonstrate the numerical behavior of the iterative substructuring methods we consider the same model problem as in Section 4.2.4 on a unit square that has been partitioned into many subdomains which are again smaller squares.

Computational Results 4.3.1: Classical Iterative Substructuring

Purpose: Demonstrate the effect of the ratio of H/h on the convergence of Algorithm 4.3.2.

PDE: Poisson problem with homogeneous Dirichlet boundary conditions.

Domain: The unit square.

Discretization: Piecewise linear finite elements on a uniform mesh.

Calculations: In the first table we present the convergence for a fixed 255×255 mesh while decreasing the number of subdomains. As the number of subdomains decreases, H increases and hence the convergence deteriorates. We list the results for two different interface preconditioners, the J operator introduced in Section 4.2.4 and the change to hierarchical basis introduced in Section 4.2.3.

Convergence for the Many Subdomain Case with Fixed Mesh

	J Operator		Hier. Basis	
Subdomains	κ	Iter.	κ	Iter.
64^2	5.37	10	6.37	12
32^2	8.69	13	10.27	15
16^2	12.72	15	14.79	17
8^2	17.30	17	19.33	18
4^2	21.90	17	23.45	19

In the next table we increase the number of subdomains as we refine the mesh; thus H/h remains constant, here $H/h = 16$.

	J Operator		Hier. Basis	
Mesh points	κ	Iter.	κ	Iter.
31^2	9.62	11	11.85	11
63^2	11.84	13	13.03	15
127^2	12.54	15	14.18	16
255^2	12.72	15	14.79	17

Discussion: In the first table one can easily see the $O(1 + \log(H/h))$ growth in the iteration counts and the $O((1+\log(H/h))^2)$ growth in the condition number as H is made larger. In the second table the growths are bounded by an asymptotic limit that they have not yet reached.

Notes and References for Section 4.3.4

- The paper by Bramble, Pasciak, and Schatz [Bra86] is considered the first paper on Schur complement methods to deal with vertex points satisfactorily.
- The observation about using a partial change to hierarchical basis first appeared in Smith [Smi90].
- The null space property was used by Bramble, Pasciak, and Schatz. See also Mandel [Man90a].
- Keyes and Gropp [Key90b], [Gro92a], [Key90a] have considered many variants of these algorithms for nonsymmetric problems.
- The values in Computational Results 4.3.1 are from Tong, Chan, and Kuo [Ton91].

4.3.5 Iterative Substructuring in Three Dimensions

The Schur complement methods for problems in two dimensions have natural extensions to three dimensions, which, unfortunately, have poorer convergence properties.

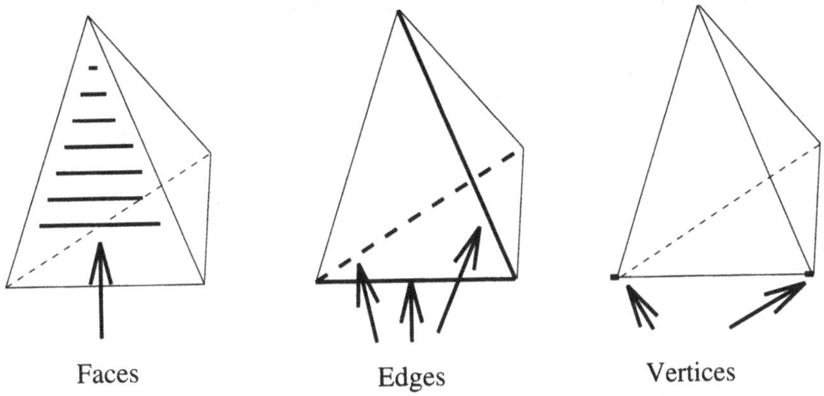

Figure 4.12. Interface, Γ, in Three Dimensions

Thus a variety of modifications to the basic iterative substructuring method have been proposed. This is an active area of research; several different approaches have been studied. In order to limit our scope, we will discuss only two algorithms that have good convergence properties. The first is the **wire basket** algorithm discussed in this section; the second method, the **vertex space** method, is discussed in the next section. Many other algorithms with similar properties and constructions are possible.

In three dimensions the interface, Γ, can be decomposed into faces, F_i, edges and vertices, see Figure 4.12. We can also decompose the vector of interface unknowns as $u_B = (u_F \, u_E \, u_V)^T$. We use the same notation as in Section 4.3.4 and partition the Schur complement similarly,

$$S = \begin{pmatrix} S_{FF} & S_{FE} & S_{FV} \\ S_{EF} & S_{EE} & S_{EV} \\ S_{VF} & S_{VE} & S_{VV} \end{pmatrix}.$$

After a partial change to hierarchical basis the Schur complement may be written as

$$S = \begin{pmatrix} I & 0 & 0 \\ 0 & I & 0 \\ -\tilde{R}_F & -\tilde{R}_E & I \end{pmatrix} \begin{pmatrix} S_{FF} & S_{FE} & \tilde{S}_{FV} \\ S_{EF} & S_{EE} & \tilde{S}_{EV} \\ \tilde{S}_{VF} & \tilde{S}_{VE} & \tilde{S}_{VV} \end{pmatrix} \begin{pmatrix} I & 0 & -\tilde{R}_F^T \\ 0 & I & -\tilde{R}_E^T \\ 0 & 0 & I \end{pmatrix}. \tag{4.10}$$

We then drop the couplings between different faces, different edges, edges and vertices, faces and vertices, and faces and edges to obtain

$$\begin{pmatrix} I & 0 & 0 \\ 0 & I & 0 \\ -\tilde{R}_F & -\tilde{R}_E & I \end{pmatrix} \begin{pmatrix} \tilde{S}_{FF} & 0 & 0 \\ 0 & \tilde{S}_{EE} & 0 \\ 0 & 0 & \tilde{S}_{VV} \end{pmatrix} \begin{pmatrix} I & 0 & -\tilde{R}_F^T \\ 0 & I & -\tilde{R}_E^T \\ 0 & 0 & I \end{pmatrix}.$$

The matrix $\tilde{S}_{FF}$ is block diagonal, with a block for each face; $\tilde{S}_{EE}$ is a block diagonal with a block for each edge; and $\tilde{S}_{VV}$ is a sparse matrix which generally couples each vertex to its neighboring vertices. Again, since we do not wish to form S explicitly we replace $\tilde{S}_{FF}$, $\tilde{S}_{EE}$, and $\tilde{S}_{VV}$ with approximations (what we call interface preconditioners) that are easier to compute, constructing our preconditioner

$$B^{-1} = \begin{pmatrix} I & 0 & 0 \\ 0 & I & 0 \\ -\tilde{R}_F & -\tilde{R}_E & I \end{pmatrix} \begin{pmatrix} \hat{S}_{FF} & 0 & 0 \\ 0 & \hat{S}_{EE} & 0 \\ 0 & 0 & A_{VV} \end{pmatrix} \begin{pmatrix} I & 0 & -\tilde{R}_F^T \\ 0 & I & -\tilde{R}_E^T \\ 0 & 0 & I \end{pmatrix}$$

or

$$B = \begin{pmatrix} I & 0 & \tilde{R}_F^T \\ 0 & I & \tilde{R}_E^T \\ 0 & 0 & I \end{pmatrix} \begin{pmatrix} \hat{S}_{FF}^{-1} & 0 & 0 \\ 0 & \hat{S}_{EE}^{-1} & 0 \\ 0 & 0 & A_{VV}^{-1} \end{pmatrix} \begin{pmatrix} I & 0 & 0 \\ 0 & I & 0 \\ \tilde{R}_F & \tilde{R}_E & I \end{pmatrix}.$$

The operator $\hat{S}_{FF}^{-1}$ is block diagonal; each block is associated with one face and could be built from any of the interface preconditioners introduced in Section 4.2. $\hat{S}_{EE}^{-1}$ can be as simple as a diagonal matrix, and A_{VV}^{-1} is a coarse grid operator possibly obtained by using linear finite elements, treating the subdomains as elements.

If $\tilde{R} = (\tilde{R}_F\ \tilde{R}_E\ I)$ and we introduce the restriction operators, $\tilde{R}_{F^i}$, for each face, and, $\tilde{R}_{E^j}$, for each edge, we can rewrite B as

$$B = \tilde{R}^T A_{VV}^{-1} \tilde{R} + \sum_i \tilde{R}_{F^i}^T B_{F^i} \tilde{R}_{F^i} + \sum_j \tilde{R}_{E^j}^T B_{E^j} \tilde{R}_{E^j}.$$

The B_{F^i} are the interface preconditioners associated with a face (i runs over all faces between substructures) while the B_{E^i} are (diagonal) edge preconditioners (for all edges in the wire basket).

It has been shown that using this type of preconditioner for certain model problems results in preconditioned systems whose condition numbers grow as fast as $O(\frac{H}{h}(1 + \log(H/h))^2)$. This result is much worse than the result in two dimensions, $O((1 + \log(H/h))^2)$. We will now demonstrate how this method may be modified so that its convergence behavior is the same as the standard iterative substructuring method in two dimensions.

We begin by observing that the iterative substructuring method in three dimensions also satisfies the null space property. As S may be assembled from its substructure contributions

$$S = \sum_{\Omega_i} \tilde{R}_i^T S^{(i)} \tilde{R}_i,$$

the preconditioner B^{-1} may be written as

$$B^{-1} = \sum_{\Omega_i} \tilde{R}_i^T B^{(i)^{-1}} \tilde{R}_i,$$

where

$$B^{(i)^{-1}} = \begin{pmatrix} I & 0 & 0 \\ 0 & I & 0 \\ -\tilde{R}_{F^i} & -\tilde{R}_{E^i} & I \end{pmatrix} \begin{pmatrix} \hat{S}_{FF}^{(i)} & 0 & 0 \\ 0 & \hat{S}_{EE}^{(i)} & 0 \\ 0 & 0 & A_{VV}^{(i)} \end{pmatrix} \begin{pmatrix} I & 0 & -\tilde{R}_{F^i}^T \\ 0 & I & -\tilde{R}_{E^i}^T \\ 0 & 0 & I \end{pmatrix}.$$

For all subdomains Ω_i, the null spaces of $S^{(i)}$ and $B^{(i)^{-1}}$ are identical. (Again, recall that in this section we restrict our attention to scalar PDEs.)

It is also possible to construct a preconditioner with the null space property, using only piecewise constant interpolation. We consider only problems where $S^{(i)}$ is symmetric and positive semidefinite.

For each substructure i define

$$\hat{S}^{(i)} = \begin{pmatrix} \hat{S}^{(i)}_{FF} & 0 & 0 \\ 0 & \hat{S}^{(i)}_{EE} & 0 \\ 0 & 0 & \tilde{S}^{(i)}_{VV} \end{pmatrix}.$$

Furthermore define $\hat{S} = \sum_{\Omega_i} \tilde{R}_i^T \hat{S}^{(i)} \tilde{R}_i$, by standard subassembly.

Define the matrix

$$M^{(i)} = \tilde{R}_i^T \left(I - \frac{z^{(i)} z^{(i)^T}}{z^{(i)^T} z^{(i)}} \right)^T \hat{S}^{(i)} \left(I - \frac{z^{(i)} z^{(i)^T}}{z^{(i)^T} z^{(i)}} \right) \tilde{R}_i.$$

For scalar problems, $z^{(i)}$ is simply the vector of all ones, $z^{(i)} = (1, 1, \ldots, 1)^T$. More generally this method can be used for coupled systems of elliptic partial differential equations. The matrix $z^{(i)}$ will then have more than one column, and the columns will span the null space of $S^{(i)}$. The term $(I - z^{(i)} z^{(i)^T} / z^{(i)^T} z^{(i)})$ is included to force $M^{(i)}$ to have the same null space as $S^{(i)}$. We now define our preconditioner $B = M^{-1}$, where M is assembled in the standard way from all the contributions $M^{(i)}$.

The solution of the resulting preconditioned problem,

$$Mu = r,$$

may be simplified by recasting the problem as a minimization problem. First write the problem as

$$\min_u \tfrac{1}{2} u^T M u - u^T r = \min_u \sum_i \tfrac{1}{2} u^T M^{(i)} u - u^T r.$$

Let $u^{(i)}$ denote the components of u associated with $\partial\Omega_i$, that is, $u^{(i)} = \tilde{R}_i u$. Then the above minimization may be written as

$$\min_u \sum_i \min_{\bar{w}^{(i)}} \tfrac{1}{2} (u^{(i)} - \bar{w}^{(i)} z^{(i)})^T \hat{S}^{(i)} (u^{(i)} - \bar{w}^{(i)} z^{(i)}) - u^T r,$$

where $\bar{w}^{(i)}$ is a scalar. We then differentiate with respect to $\bar{w}^{(i)}$ and u, to obtain the linear system

$$z^{(i)^T} \hat{S}^{(i)} (u^{(i)} - \bar{w}^{(i)} z^{(i)}) = 0 \quad \forall i$$

$$\hat{S} u - \sum_i \bar{w}^{(i)} \tilde{R}_i^T \hat{S}^{(i)} z^{(i)} = r. \tag{4.11}$$

We then eliminate u and get the following system for the $\bar{w}^{(i)}$:

$$(z^{(i)^T}\hat{S}^{(i)}z^{(i)})\bar{w}^{(i)} - z^{(i)^T}\hat{S}^{(i)}\tilde{R}_i\hat{S}^{-1}\sum_j \bar{w}^{(j)}\tilde{R}_j^T\hat{S}^{(j)}z^{(j)} = z^{(i)^T}\hat{S}^{(i)}\tilde{R}_i\hat{S}^{-1}r.$$

This is a sparse linear system with as many unknowns as there are substructures. Once the $\bar{w}^{(i)}$ are known, u can be found easily by solving (4.11).

We can interpret this preconditioner as a two step method. It is *not* possible to solve the "coarse" problem and local problems in parallel while preserving both the null space and the convergence properties.

This coarse grid correction is very closely related to the coarse grid correction introduced for the balancing Neumann-Neumann method in Section 4.3.3. In both cases, averages of u over $\partial\Omega_i$ are calculated before the local solvers are applied.

The method may be modified into a **wire basket** based algorithm which can be completely parallelized. This method has been developed only for scalar elliptic PDEs. We first group the vertex nodes with the edge nodes into one set called the wire basket. $S^{(i)}$ may then be written as

$$S^{(i)} = \begin{pmatrix} S_{FF}^{(i)} & S_{FW}^{(i)} \\ S_{WF}^{(i)} & S_{WW}^{(i)} \end{pmatrix}.$$

Let $T^{(i)^T}$ map the average of the values of the boundary nodes of each face (the adjacent edges and vertices) onto the nodes on the corresponding face. Then $S^{(i)}$ may be written as

$$S^{(i)} = \begin{pmatrix} I & 0 \\ -T^{(i)} & I \end{pmatrix}\begin{pmatrix} S_{FF}^{(i)} & \tilde{S}_{FW}^{(i)} \\ \tilde{S}_{WF}^{(i)} & \tilde{S}_{WW}^{(i)} \end{pmatrix}\begin{pmatrix} I & -T^{(i)^T} \\ 0 & I \end{pmatrix}.$$

Note the similarity to (4.10), except now rather than working with a linear partial hierarchical basis, we are working with a piecewise constant partial hierarchical basis. Now, as with the standard iterative substructuring method, drop the couplings between the various faces, and the faces and the wire basket. Also replace $\tilde{S}_{WW}^{(i)}$ by $G^{(i)}$, introduced below, to obtain

$$B^{(i)^{-1}} = \begin{pmatrix} I & 0 \\ -T^{(i)} & I \end{pmatrix}\begin{pmatrix} \hat{S}_{FF}^{(i)} & 0 \\ 0 & G^{(i)} \end{pmatrix}\begin{pmatrix} I & -T^{(i)^T} \\ 0 & I \end{pmatrix}.$$

If the null space of $G^{(i)}$ is the same as that of $\tilde{S}_{WW}^{(i)}$, then $\mathrm{Null}(B^{(i)}) = \mathrm{Null}(S^{(i)})$ and we preserve the null space property. Since, for scalar problems, the null space is the space of constant functions, we can apply the previously used trick to define G,

$$v^T G^{(i)} v = \min_{\bar{w}_i}(v - \bar{w}_i z^{(i)})^T (v - \bar{w}_i z^{(i)}).$$

Define $B^{-1} = \sum_{\Omega_i} \tilde{R}_i^T B^{(i)^{-1}} \tilde{R}_i$; then B may be written as

$$B = \begin{pmatrix} I & T^T \\ 0 & I \end{pmatrix}\begin{pmatrix} \hat{S}_{FF}^{-1} & 0 \\ 0 & G^{-1} \end{pmatrix}\begin{pmatrix} I & 0 \\ T & I \end{pmatrix},$$

or

$$B = \sum_{\Omega_i} \tilde{R}_{F^i}^T B_{F^i} \tilde{R}_{F^i} + \begin{pmatrix} T^T \\ I \end{pmatrix} G^{-1} \begin{pmatrix} T & I \end{pmatrix}.$$

The action of G^{-1} may be applied by solving a minimization problem similar to that used above for the application of M^{-1}. In this procedure a separate average for each face shared by two subdomains is calculated.

The wire basket algorithm has a convergence rate independent of the number of subdomains and independent of the jumps in the coefficients of the PDE between subdomains. The condition number of the resulting preconditioned system grows like $O((1 + \log(H/h))^2)$ for certain model problems in three dimensions.

Notes and References for Section 4.3.5

- The paper by Bramble, Pasciak, and Schatz [Bra89] is the first to deal with Schur complement methods in three dimensions with $O((1 + \log(H/h))^2)$ condition numbers.
- Dryja [Dry88] then used the same approach to derive a Neumann-Dirichlet algorithm whose convergence did not depend on the number of subdomains.
- Mandel [Man90a] provided a simple, linear algebra based explanation for why the Bramble, Pasciak, and Schatz algorithm has the null space property. Smith [Smi91] later modified Mandel's technique to produce the wire basket algorithm presented in this section.
- Some numerical experiences for the wire basket method may be found in Smith [Smi93].
- In order to obtain the condition number bound $O((1+\log(H/h))^2)$ for the wire basket method, Smith had to assume a certain scaling of the coarse grid problem. In practice, this scaling is of no importance.
- For a much more complete discussion of many iterative substructuring algorithms, for problems in three dimensions, see Dryja, Smith, and Widlund [Dry94a].

4.3.6 Iterative Substructuring with Explicit Overlap

Most iterative substructuring algorithms are characterized by the $O((1 + \log(H/h))^2)$ growth in the condition number as the mesh is refined. The overlapping Schwarz methods discussed in Chapters 1–3 exhibit no such growth. This difference can be tracked down to the fact that the vertex points (in two dimensions) or wire basket points (in three dimensions) are not involved in any of the local solvers, only in the coarse grid solve. See Section 5.3.2, where an explicit derivation of the $O((1 + \log(H/h))^2)$ term is developed for a class of model problems in two dimensions solved using iterative substructuring.

By introducing additional local solvers associated with the points near each vertex point (in two dimensions) and near each vertex and each edge (in three dimensions) it is possible to convert the standard iterative substructuring algorithm into a method with a condition number, and hence convergence rate, that does not grow as the mesh is refined. This method is known as the **vertex space** method.

Computational Results 4.3.2: Vertex Space Method

Purpose: To demonstrate the effect of including the "vertex" spaces on the convergence of the vertex space method.

PDE: The equations of linear elasticity in two dimensions.

Domain: The unit square.

Discretization: Quadrilateral Q_1 elements.

Calculations: In the first table the overlap was kept as a fixed percentage of H. We list the condition number and number of iterations, with and without the "vertex" spaces.

No. of Subdomains	Nodes along Edge	No. of Unknowns on Γ	No Preconditioner		Without "Vertex" Spaces		With "Vertex" Spaces	
16	3	162	22.16	19	10.52	15	3.51	10
	7	354	46.63	28	14.84	17	3.51	10
	15	738	96.34	41	19.83	19	3.56	10
	31	1506	196	60	25.51	20	3.62	10
64	3	770	84.82	37	12.13	18	3.85	10
	7	1666	178	55	16.92	19	3.85	10
	15	3458	368	79	22.37	22	3.84	10
	31	7042	747	>100	28.52	25	3.89	10
256	3	3330	334	75	12.42	18	3.91	10
	7	7170	705	>100	17.31	19	3.90	10
	15	14850	1453	>100	22.88	22	3.89	10
	31	30210	2921	>100	29.15	25	3.94	10

Below (see Figure 4.13), the overlap is increased from 0 (no vertex space problem) to demonstrate the effect of the overlap. Only a small overlap is needed in practice.

Overlap in nodes	0	1	2	3	4	5	6	7	8
Condition number	28.55	5.59	4.85	4.36	4.01	3.89	3.88	3.89	3.89
Iterations	25	12	11	10	10	10	10	10	10

Discussion: Details may be found in Smith [Smi92].

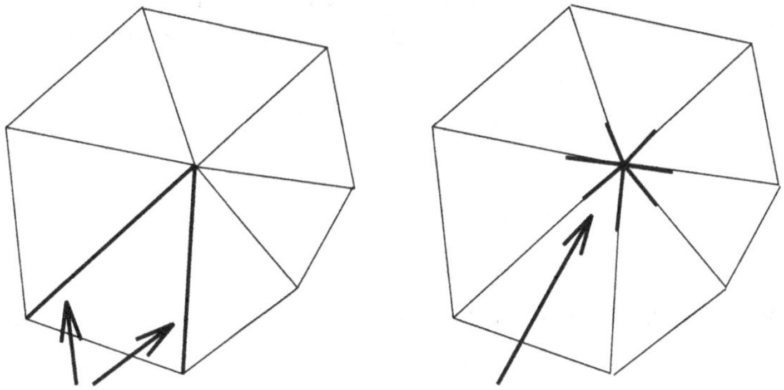

Figure 4.13. Vertex Space Overlap in Two Dimensions

Notes and References for Section 4.3.6

- A similar method, for a few subdomains, was proposed and analyzed by Nepomnyaschikh [Nep86].
- The vertex space method is often called the Copper Mountain algorithm since Smith [Smi92] first introduced it, independently of Nepomnyaschikh, at a Copper Mountain conference on iterative methods held in 1990.
- Chan and Mathew [Cha91b] have considered using probing and extensions of the J operator, to calculate approximations for the local vertex solvers. See also Chan, Mathew, and Shao [Cha94b].

4.4 Inexact Subdomain Solvers

In the previous sections we have assumed that the interior problems were solved exactly, to machine precision. This is an expensive computational procedure, especially since it must be done at each iteration. In this section we show how the Schur complement algorithms may be modified to use approximate interior solves. Recall that A^{-1} can be written as

$$A^{-1} = \begin{pmatrix} I & -A_{II}^{-1}A_{IB} \\ 0 & I \end{pmatrix} \begin{pmatrix} A_{II}^{-1} & 0 \\ 0 & S^{-1} \end{pmatrix} \begin{pmatrix} I & 0 \\ -A_{BI}A_{II}^{-1} & I \end{pmatrix}. \tag{4.12}$$

So if we have a good preconditioner for S, call it B_S, and a good preconditioner for A_{II}, that is, an approximate solver for the interior problems called B_I, then we can create a preconditioner for A of the form

$$B = \begin{pmatrix} I & -B_I A_{IB} \\ 0 & I \end{pmatrix} \begin{pmatrix} B_I & 0 \\ 0 & B_S \end{pmatrix} \begin{pmatrix} I & 0 \\ -A_{BI}B_I & I \end{pmatrix}.$$

Note that an application of B on a vector only requires two applications of B_I and one application of B_S. For nonsymmetric problems, where the symmetry of B is not important, one could drop the first or third term from B. Again we note that none of these matrices need be formed explicitly. It is also possible to use different approximate interior solvers in the three terms of B. For instance, for nonsymmetric problems one may use

$$B = \begin{pmatrix} I & -B_I^{(L)}A_{IB} \\ 0 & I \end{pmatrix} \begin{pmatrix} B_I^{(C)} & 0 \\ 0 & B_S \end{pmatrix} \begin{pmatrix} I & 0 \\ -A_{BI}B_I^{(R)} & I \end{pmatrix}.$$

Computational Results 4.4.1: Wire Basket Method with Approximate Solvers

Purpose: Demonstrate the effect of using approximate solvers to solve the subdomain problems for the wire basket algorithm.

PDE: Poisson problem.

Domain: The unit cube.

Growth in Condition Numbers for 64 Subdomains ($H = \frac{1}{4}$)

	Total			$\kappa(BA)$	
H/h	Unknowns	$\kappa(A)$	$\kappa(S)$	Exact	Approx.
4	3,375	103	53.8	9.4	9.4
8	29,791	414	122	15.0	15.0
12	103,823	933	192	20.1	20.1
16	250,047	1,656	261	24.2	24.4
20	493,039	2,593	331	27.9	28.1
24	857,375	3,734	401	31.1	31.3
28	1,367,631	5,083	> 500	33.9	34.2
32	2,048,383	6,640	> 500	36.5	36.9
Observed growth:		$(1/h)^2$	$1/(Hh)$	$(1 + \log(H/h))^2$	

Growth in Condition Numbers for 216 Subdomains ($H = \frac{1}{6}$)

	Total			$\kappa(BA)$	
H/h	Unknowns	$\kappa(A)$	$\kappa(S)$	Exact	Approx.
4	12,167	232	119	9.9	9.9
8	103,823	933	269	16.1	16.1
12	357,911	2,099	421	21.5	21.5
16	857,375	3,734	573	26.0	26.1
20	1,685,159	5,835	726	29.8	30.0
Observed growth:		$(1/h)^2$	$1/(Hh)$	$(1 + \log(H/h))^2$	

Discretization: Centered finite differences.

Calculations: We have used both an exact subdomain solve and an approximate subdomain solve calculated by using one V-cycle of classical multigrid.

Discussion: As one may see, the effect of the approximate solve is barely visible. Full details may be found in Smith [Smi93].

There is an important difference between iterative substructuring algorithms and overlapping Schwarz methods related to using approximate solvers. With overlapping Schwarz methods when the local subdomain solvers are replaced by a spectrally equivalent operator (solver) one always retains the same overall asymptotic convergence behavior. This is not the case for iterative substructuring. If one replaces the A_{II}^{-1} that appears in the first and third terms in (4.12) with a spectrally equivalent solver the resulting preconditioner is *not* necessarily spectrally equivalent to A.

4.4.1 "Optimal" Iterative Substructuring Methods

Schur complement algorithms are generally not optimal, in that the amount of work grows more than linearly with the number of unknowns, since the number of iterations increases as a result of the increase in the condition number. There is a clever trick that may be applied to certain problems to obtain optimal preconditioners using Schur complement approaches. For simplicity we consider only the case where A is symmetric, positive definite.

Let N denote the total number of unknowns while $n \ll N$ denotes the number of unknowns on the interface. The idea is first to construct an operator $\tilde{B}_S$ that is spectrally equivalent to S^{-1} but which may be applied in $O(n\log(n))$ operations. Usually this is done by fast sine transforms, see Section 4.2.4. Now the action of $\tilde{B}_S$ may be calculated by solving a linear system with the preconditioned conjugate gradient method using the preconditioner B_S. Since only $O(1+\log(H/h))$ iterations are required for this inner iteration, the total operation count to apply $\tilde{B}_S$ to a vector can be made less than $O(N)$. Thus we have a preconditioner for S that is spectrally equivalent to S and can be applied with $O(N)$ work. This approach may be used with any number of subdomains.

Notes and References for Section 4.4

- There has yet to be developed a complete, mathematically satisfactory approach to deal with the use of approximate solvers in iterative substructuring. See the discussion in Haase, Langer, and Meyer [Haa90].
- See, for example, Khoromskij and Wendland [Kho92], where the optimal iterative substructuring approach is applied to the boundary element method.

4.5 Implementation Issues

Making a good choice of an iterative substructuring method for a particular problem is often difficult. This is because each method has its own advantages and liabilities.

For simple, scalar second order diffusion dominated elliptic PDEs, with constant or slowly varying coefficients, the standard iterative substructuring method introduced in Section 4.3.5 performs well in two dimensions. The J operator introduced in Section 4.2.4 performs well as the interface preconditioner. For three dimensions the algorithm introduced in Section 4.3.4 should be replaced with the wire basket algorithm introduced at the end of Section 4.3.5.

For second order elliptic PDEs with piecewise smooth coefficients (assuming that the jumps in the coefficients occur only on the interface between coarse grid elements) the above recommendation still holds.

The balancing Neumann-Neumann method, Section 4.3.3, appears to be robust and can be applied equally to scalar and multicomponent PDEs as well as to unstructured grids. It is, however, an expensive preconditioner since at each iteration a Dirichlet and a Neumann problem must be solved on each subdomain. In addition the coarse grid correction matrix must be explicitly constructed.

A nice feature of the balancing algorithm is that a single code may be written that can be reused for many different PDEs and discretizations.

Notes and References for Section 4.5

- Several extensive numerical studies of iterative substructuring type algorithms for problems in two dimensions have been performed by Gropp and Keyes [Key87], [Key92b], [Gro91], and X. Cai, Gropp, and Keyes [Cai92a], [Cai94a].
- In three dimensions, the wire basket method has been implemented by Smith [Smi93].
- A sequential version of the balancing Neumann-Neumann method, written by Mandel, is available from MG-Net.

4.6 Variational Formulation

The variational interpretation of Schur complement methods is more subtle than that for the overlapping Schwarz methods, but the fundamental ideas are the same.

Consider the space V_B^h, which is the restriction of all functions in V^h to Γ, the interface between the subdomains. A basis for this space is given by $\{\phi_{B_i}\}$, where ϕ_{B_i} are the finite element basis functions ϕ_i restricted to the interface. The "extra" ϕ_k, $\{\phi_{I_i}\}$ associated with nodes in the interior of Ω_k are zero on Γ and are not included. Any function $\boldsymbol{u}_B \in V_B^h$ can be represented as

$$\boldsymbol{u}_B = \sum_i u_{B_i} \phi_{B_i}.$$

We introduce a bilinear form on V_B^h by

$$s(\boldsymbol{u}_B, \boldsymbol{v}_B) = u_B^T S v_B.$$

The linear system $Su_B = g$ can be rewritten as

$$s(\boldsymbol{u}_B, \boldsymbol{v}_B) = (\boldsymbol{v}_B, \boldsymbol{g})_{l^2} \qquad \forall \boldsymbol{v}_B \in V_B^h,$$

where $\boldsymbol{g} = \sum_i g_i \phi_{B_i}$ and $(\boldsymbol{v}_B, \boldsymbol{g})_{l^2}$ is the discrete l^2 inner product $v_B^T g$.

Define the discrete harmonic extension of $\boldsymbol{u}_B$ to be $\boldsymbol{u}_{\mathcal{H}} = \sum_i (u_B)_i \phi_{B_i} + \sum_j (u_I)_j \phi_{I_j}$, where $u_I = -A_{II}^{-1} A_{IB} u_B$. Then, it follows that

$$s(\boldsymbol{u}_B, \boldsymbol{u}_B) = a(\boldsymbol{u}_{\mathcal{H}}, \boldsymbol{u}_{\mathcal{H}}).$$

The space of all $\boldsymbol{u}_{\mathcal{H}}$, $\tilde{V}^h$, is the subspace of V^h of **discrete harmonic functions**. When $a(\cdot,\cdot)$ and hence $s(\cdot,\cdot)$ are symmetric, positive definite, then the bilinear form $s(\cdot,\cdot)$ has the minimization property,

$$s(\boldsymbol{u}_B, \boldsymbol{u}_B) = \min_{\substack{\boldsymbol{u} \in V^h \\ \boldsymbol{u}|_\Gamma = \boldsymbol{u}_B}} a(\boldsymbol{u}, \boldsymbol{u}).$$

This may be expressed in matrix notation as

$$u_B^T S u_B = \min_{u_I} (u_I^T u_B^T) A \begin{pmatrix} u_I \\ u_B \end{pmatrix}.$$

We see that the discrete harmonic extension is the energy minimizing extension. Also observe that for the symmetric positive definite case the spaces $\tilde{V}^h$ and $V_i^h = H_0^1(\Omega_i) \cap V^h$ are orthogonal in the $a(\cdot,\cdot)$ inner product. This follows since, for $\boldsymbol{v} \in V_i^h$ and $\boldsymbol{u} \in \tilde{V}^h$,

$$a(\boldsymbol{v}, \boldsymbol{u}) = (v_I^T\ 0) \begin{pmatrix} A_{II} & A_{IB} \\ A_{IB}^T & A_{BB} \end{pmatrix} \begin{pmatrix} -A_{II}^{-1} A_{IB} u_B \\ u_B \end{pmatrix} = 0.$$

For simplicity, in the rest of this section, we restrict ourselves to two dimensions and piecewise linear finite elements. A similar construction is possible in three dimensions.

We will demonstrate that by a careful choice of the overlapping subdomains it is possible to construct the standard iterative substructuring algorithm as a two level overlapping Schwarz method. The main reason for this construction is to provide insight into the convergence of iterative substructuring algorithms (this is done in Section 5.3.2) and to determine commonalities between various domain decomposition methods.

Let E_{ij} denote the common edge shared by subdomains Ω_i and Ω_j. E_{ij} is an open line segment that does not contain its endpoints. Let $\tilde{V}_{ij}^h$ be the subspace of $\tilde{V}^h$ consisting of all functions that are zero on $\Gamma \setminus E_{ij}$.

Let V^H denote the subspace of functions in V^h that are continuous and piecewise linear on the substructures. Note that since the functions in V^H are linear on each substructure, they are discrete harmonic. That is, $V^H \subset \tilde{V}^h$.

Given this proliferation of subspaces it is possible to show that

$$\begin{aligned} V^h &= \bigoplus_i V_i^h \oplus \tilde{V}^h, \\ &= \bigoplus_i V_i^h \oplus V^H \bigoplus_{ij} \tilde{V}_{ij}^h. \end{aligned}$$

For each subspace we define the a-projection of $\boldsymbol{u} \in V^h$ onto that subspace. Define $\boldsymbol{P}_i\boldsymbol{u} \in V_i^h$ by

$$a(\boldsymbol{P}_i\boldsymbol{u}, \boldsymbol{v}) = a(\boldsymbol{u}, \boldsymbol{v}) \qquad \forall \boldsymbol{v} \in V_i^h.$$

Similarly $\boldsymbol{P}_0\boldsymbol{u} \in V^H$ is given by

$$a(\boldsymbol{P}_0\boldsymbol{u}, \boldsymbol{v}) = a(\boldsymbol{u}, \boldsymbol{v}) \qquad \forall \boldsymbol{v} \in V^H.$$

Finally, define $\boldsymbol{P}_{ij}\boldsymbol{u} \in \tilde{V}_{ij}^h$ by

$$a(\boldsymbol{P}_{ij}\boldsymbol{u}, \boldsymbol{v}) = a(\boldsymbol{u}, \boldsymbol{v}) \qquad \forall \boldsymbol{v} \in \tilde{V}_{ij}^h.$$

From Sections 1.7 and 2.7 we know the matrix representation of the projections $\boldsymbol{P}_i$ and $\boldsymbol{P}_0$

$$P_i = R_i^T A_{\Omega_i}^{-1} R_i A$$

and

$$P_0 = R^T A_C^{-1} R A.$$

As in Chapter 2, R^T denotes (linear) interpolation from the coefficients on the coarse grid to the fine grid. Also A_{Ω_i} is the submatrix of A associated with the (in this section nonoverlapping) subdomain Ω_i. The matrix A_C is the stiffness matrix on the coarse grid defined by the substructures.

We now derive the matrix form of the "interface" projection $\boldsymbol{P}_{ij}$. Since $P_{ij}u$ and v are the coefficients associated with functions in $\tilde{V}_{ij}^h$, we know that they can be uniquely represented as

$$P_{ij}u = \begin{pmatrix} -A_{II}^{-1}A_{IB} \\ I \end{pmatrix} \tilde{R}_{ij}^T \bar{u}$$

and

$$v = \begin{pmatrix} -A_{II}^{-1}A_{IB} \\ I \end{pmatrix} \tilde{R}_{ij}^T \bar{v},$$

where $\bar{u}$ and $\bar{v}$ are vectors with one element for each node on E_{ij}. The restriction matrix $\tilde{R}_{ij}$ is as defined in Section 4.3.4. From the definition of $\boldsymbol{P}_{ij}$ given above, for all $\bar{v}$,

$$\bar{v}^T \tilde{R}_{ij}(-A_{IB}^T A_{II}^{-1}\ I) \begin{pmatrix} A_{II} & A_{IB} \\ A_{IB}^T & A_{BB} \end{pmatrix} \begin{pmatrix} -A_{II}^{-1}A_{IB} \\ I \end{pmatrix} \tilde{R}_{ij}^T \bar{u} = \bar{v}^T \tilde{R}_{ij}(-A_{IB}^T A_{II}^{-1}\ I)Au.$$

Hence,

$$\bar{u} = (\tilde{R}_{ij} S \tilde{R}_{ij}^T)^{-1} \tilde{R}_{ij}(-A_{IB}^T A_{II}^{-1}\ I)Au.$$

Thus, with $S_{ij} = \tilde{R}_{ij} S \tilde{R}_{ij}^T$,

$$P_{ij}u = \begin{pmatrix} -A_{II}^{-1}A_{IB} \\ I \end{pmatrix} \tilde{R}_{ij}^T S_{ij}^{-1} \tilde{R}_{ij}(-A_{IB}^T A_{II}^{-1}\ I)Au.$$

The matrix S_{ij} is merely the submatrix of S associated with the edge E_{ij}.

Adding up all of the projection matrices results in

$$P_0 + \sum_{\Omega_i} P_i + \sum_{E_{ij}} P_{ij} = \begin{pmatrix} I & -A_{II}^{-1}A_{IB} \\ 0 & I \end{pmatrix} \begin{pmatrix} A_{II}^{-1} & 0 \\ 0 & B_S \end{pmatrix} \begin{pmatrix} I & 0 \\ -A_{BI}A_{II}^{-1} & I \end{pmatrix} A,$$

where

$$B_S = \tilde{R}^T A_C^{-1} \tilde{R} + \sum_{E_{ij}} \tilde{R}_{ij}^T S_{ij}^{-1} \tilde{R}_{ij}$$

or, equivalently,

$$B_S = \begin{pmatrix} I & \tilde{R}_E^T \\ 0 & I \end{pmatrix} \begin{pmatrix} \tilde{S}_{EE}^{-1} & 0 \\ 0 & A_C^{-1} \end{pmatrix} \begin{pmatrix} I & 0 \\ \tilde{R}_E & I \end{pmatrix}.$$

Thus we can conclude that the standard iterative substructuring preconditioner in two dimensions involves the projection of the error onto a particular set of subspaces of V^h. This result is used in Section 5.3.2 to prove convergence of the standard iterative substructuring method.

It is possible also to view the iterative substructuring algorithms as Schwarz methods on the space $\tilde{V}^h$ with bilinear form $s(\cdot,\cdot)$ rather than as Schwarz methods on the entire space V^h. This is because the subspaces V_i^h are orthogonal to $\tilde{V}^h$ so, in fact, the linear system completely decouples into local subdomain problems and a Schur complement problem on Γ. It only appears not to decouple because the solution of the Schur complement problem requires the solution of subdomain problems.

We conclude this chapter by showing that the Neumann-Neumann algorithm may also be viewed as a Schwarz method. Define the subdomain boundary spaces by

$$\tilde{V}_i^h = \{\boldsymbol{u}^h \in \tilde{V}^h | \boldsymbol{u}^h(x_k) = 0, \ \forall x_k \in \cup_j \partial\Omega_j \setminus \partial\Omega_i\}.$$

The x_k denote the finite element nodes on the boundary of all substructures excluding Ω_i. The subspace decomposition used in the one level Neumann-Neumann method is

$$V^h = \bigoplus_i V_i^h \oplus \bigoplus_i \tilde{V}_i^h.$$

For the interior subdomains (that is, all subdomains that do not have an associated Dirichlet boundary condition) we need an auxiliary bilinear form,

$$\hat{a}_{\Omega_i}(\boldsymbol{u}, \boldsymbol{v}) = a_{\Omega_i}(\boldsymbol{u}, \boldsymbol{v}) + H^{-2} \int_{\Omega_i} uv \, dx.$$

The second term is needed to ensure that the local Schur complement associated with Ω_i is nonsingular (see Section 4.3.1, where this is expressed by matrix notation as the Schur complement of the matrix $A_i + \alpha I$). For subdomains that have an associated Dirichlet boundary condition the second term is not needed. Let $\hat{\mathcal{H}}_i \boldsymbol{u}$ denote the discrete harmonic extension of $\boldsymbol{u}$ into the interior of Ω_i using the $\hat{a}_{\Omega_i}(\cdot,\cdot)$ bilinear form. Also let λ_i scale the value of $\boldsymbol{u}$ at each finite element node by the number of subdomains it is contained in. Define an additional bilinear form by

$$a_i(\boldsymbol{u}, \boldsymbol{v}) = \hat{a}_{\Omega_i}(\hat{\mathcal{H}}_i I^h \lambda_i \boldsymbol{u}, \hat{\mathcal{H}}_i I^h \lambda_i \boldsymbol{v}).$$

Finally we define the projectionlike operator $\tilde{\boldsymbol{P}}_i \boldsymbol{u} \in \tilde{V}_i^h$ by

$$a_i(\tilde{\boldsymbol{P}}_i \boldsymbol{u}, \boldsymbol{v}) = a(\boldsymbol{u}, \boldsymbol{v}) \qquad \forall \boldsymbol{v} \in \tilde{V}_i^h. \tag{4.13}$$

The matrix representation of $\tilde{\boldsymbol{P}}_i$ is given by

$$\tilde{P}_i u = \begin{pmatrix} -A_{II}^{-1} A_{IB} \\ I \end{pmatrix} D \tilde{R}_i^T S^{(i)^{-1}} \tilde{R}_i D \left(-A_{IB}^T A_{II}^{-1} I \right) Au.$$

The diagonal scaling matrix D is the same as that introduced in Section 4.3.2 and was induced by including the λ_i term in the definition of $a_i(\cdot, \cdot)$.

This elaborate construction was carried out to demonstrate that the corrections calculated in the Neumann-Neumann preconditioner are "projections" on $\tilde{V}_i^h$ using the auxiliary bilinear form $a_i(\cdot, \cdot)$.

Notes and References for Section 4.6

- The use of the decomposition into $H_0^1(\Omega_i) \cap V^h$ and the space of discrete harmonic functions has been understood for many years; see, for instance, Matsokin and Nepomnyaschikh [Mats85] and Nepomnyaschikh [Nep84], [Nep86].
- The decomposition was later used by Bramble, Pasciak and Schatz [Bra86], [Bra87], [Bra88], and [Bra89].
- Dryja and Widlund [Dry89] introduced the authors to the Schwarz analysis approach to iterative substructuring methods.
- Dryja and Widlund [Dry90], [Dry95] also were the first to formulate the Neumann-Neumann algorithms as Schwarz methods.

5

A Convergence Theory

OVER THE PAST ten years a powerful framework has been developed for the design and analysis of domain decomposition and multilevel/multigrid algorithms. For linear elliptic PDEs, discretized with Galerkin finite elements, seemingly very different algorithms have important common mathematical structure. We refer to the theory of these methods as the Schwarz framework, for the related work of Schwarz in the nineteenth century. We will refer to all methods that may be analyzed by using these techniques as Schwarz methods.

Convergence analysis of domain decomposition and multigrid methods is usually quite involved. Hence it is useful to divide the analysis up into two stages. The **abstract convergence theory** consists of several basic assumptions on the subspaces arising from the algorithm plus some (often complicated) algebraic manipulation. A convergence proof for a particular algorithm can then be constructed by verifying these assumptions and applying the abstract theory.

5.1 Schwarz Framework

This section introduces the abstract framework and derives several important consequences. Most of the manipulations are algebraic, though somewhat complex, and use nothing more profound than the Cauchy-Schwarz inequality. We remind the reader that we reserve boldface (i.e., $\boldsymbol{u}$, $\boldsymbol{A}$) for functions and operators, while italics (i.e., u, A) represent discrete functions (vectors) and matrices.

The abstract theory is developed by using several basic ideas from functional analysis. We formulate the framework in this abstract manner in order to apply it in a variety of situations. Assume V is a finite dimensional Hilbert space (vector space with inner product and corresponding norm) in which we wish to solve our problem. Find $\boldsymbol{u}^* \in V$ such that

$$a(\boldsymbol{u}^*, \boldsymbol{v}) = (\boldsymbol{f}, \boldsymbol{v}) \qquad \forall \boldsymbol{v} \in V, \tag{5.1}$$

where $a(\cdot, \cdot)$ is a bilinear form and $(\cdot, \cdot)$ is the inner product on the space. Furthermore, assume that $a(\boldsymbol{v}, \boldsymbol{v}) = 0$ if and only if $\boldsymbol{v} = 0$. This problem, (5.1), is well posed by the assumptions on V and $a(\cdot, \cdot)$. The dimension of V may be very large. The conforming finite element discretization of second order, uniformly elliptic partial differential equations, for instance, leads to problems of exactly this form.

Define the linear operator $\boldsymbol{A}$ by

$$(\boldsymbol{A}\boldsymbol{u}, \boldsymbol{v}) = a(\boldsymbol{u}, \boldsymbol{v}) \qquad \forall \boldsymbol{v} \in V.$$

Then (5.1) is equivalent to the linear operator equation

$$\boldsymbol{A}\boldsymbol{u}^* = \boldsymbol{f}. \tag{5.2}$$

Once a basis has been chosen for the space V, the operator equation can be rewritten as a linear system, $Au = f$. In this chapter we will usually express the linear algebra "basis free" and write the analysis by using operators, rather than matrices.

A concrete example of a Hilbert space V and the bilinear form $a(\cdot, \cdot)$ is given in Section 1.6.

Schwarz methods solve the operator equation (5.2) iteratively by solving a sequence of subproblems defined by subspaces. Let V_i, $i = 0, \ldots, p$, be a set of auxiliary spaces, generally of much smaller dimension than V. In addition, assume that there exists a set of interpolation operators $\boldsymbol{I}_i : V_i \to V$. We will refer to the case when $V_i \subset V$ as the **nested** case. In this case the $\boldsymbol{I}_i$ are simply imbedding operators. We also assume that on each space V_i there is a bilinear form, $a_i(\cdot, \cdot) \geq 0$ (which may be $a(\cdot, \cdot)$ itself), that "approximates" $a(\cdot, \cdot)$ on V_i. In addition, we require that, for $\boldsymbol{v} \in V_i$, $a_i(\boldsymbol{v}, \boldsymbol{v}) = 0$ implies $\boldsymbol{v} = 0$.

Define the **projectionlike** operator $\tilde{\boldsymbol{T}}_i : V \to V_i$ by

$$a_i(\tilde{\boldsymbol{T}}_i \boldsymbol{u}, \boldsymbol{v}) = a(\boldsymbol{u}, \boldsymbol{I}_i \boldsymbol{v}) \qquad \forall \boldsymbol{v} \in V_i. \tag{5.3}$$

The definition of $\tilde{\boldsymbol{T}}_i$ is well posed since V_i is a finite dimensional Hilbert space. The effect of $\tilde{\boldsymbol{T}}_i$ can be calculated by solving a linear system whose dimension is the same as the dimension of V_i. To see this, let $\{\psi_k\}$ form a basis for V_i and express $\tilde{\boldsymbol{T}}_i \boldsymbol{u} = \sum_k w_k \psi_k$. Then

$$\begin{aligned} a_i(\tilde{\boldsymbol{T}}_i \boldsymbol{u}, \boldsymbol{v}) &= a(\boldsymbol{u}, \boldsymbol{I}_i \boldsymbol{v}) && \forall \boldsymbol{v} \in V_i, \\ \sum_k w_k a_i(\psi_k, \boldsymbol{v}) &= a(\boldsymbol{u}, \boldsymbol{I}_i \boldsymbol{v}) && \forall \boldsymbol{v} \in V_i, \\ \sum_k w_k a_i(\psi_k, \psi_l) &= a(\boldsymbol{u}, \boldsymbol{I}_i \psi_l) && \forall \psi_l. \end{aligned}$$

But this is simply the linear system,

$$A_i w = b^i, \tag{5.4}$$

where $(A_i)_{lk} = a_i(\psi_k, \psi_l)$ and the components of b^i are $b^i_l = a(\boldsymbol{u}, \boldsymbol{I}_i \psi_l)$. Thus, we have shown that for any given function $\boldsymbol{u} \in V$, we can compute $\tilde{\boldsymbol{T}}_i \boldsymbol{u}$ by first computing the right hand side vector b^i, then solving (5.4) for w and finally compute $\tilde{\boldsymbol{T}}_i \boldsymbol{u} = \sum_k w_k \psi_k$.

Now define the operator $\boldsymbol{T}_i = \boldsymbol{I}_i \tilde{\boldsymbol{T}}_i : V \to V$. In Section 2.8.2, we derived the matrix form of the operators $\boldsymbol{T}_i$,

$$T_i = R_i^T A_i^{-1} R_i A.$$

In the nested (Galerkin) case $A_i = R_i A R_i^T$. The matrix R_i^T is the discrete representation of the interpolation (or imbedding) from V_i to V.

It is possible to avoid the entire issue of nonnested spaces by simply redefining $V_i = \boldsymbol{I}_i V_i \subset V$. This is referred to as Method 2 in Section 2.8.2. The stiffness matrices from the subproblems are different in the two approaches. We do not adopt this approach in this book.

Writing $\tilde{\boldsymbol{T}}_i \boldsymbol{u}^* = \sum_k w_k \psi_k$ as above we obtain

$$\sum_k w_k a_i(\psi_k, \psi_l) = a(\boldsymbol{u}^*, \boldsymbol{I}_i \psi_l) = (f, \boldsymbol{I}_i \psi_l) \qquad \forall \psi_l \in V_i,$$

where we have used (5.1). By solving (5.4) for w with $b_l^i = (f, \boldsymbol{I}_i \psi_l)$ we get

$$\boldsymbol{T}_i \boldsymbol{u}^* = \boldsymbol{I}_i \tilde{\boldsymbol{T}}_i \boldsymbol{u}^* = \sum_k w_k \boldsymbol{I}_i \psi_k.$$

We have just shown that we can calculate $\boldsymbol{T}_i \boldsymbol{u}^*$ without knowing the solution $\boldsymbol{u}^*$.

In addition, proceeding exactly as above, we can calculate $\boldsymbol{T}_i \boldsymbol{u}$ for any given function $\boldsymbol{u} \in V$, by solving (5.4). Comparing with Section 2.8.2, we may express the same terms by using matrix notation. The expression $b_l^i = (f, \boldsymbol{I}_i \psi_l)$ corresponds to $b^i = R_i f$ and the calculation of $\boldsymbol{T}_i \boldsymbol{u}$ can be expressed as $T_i u = R_i^T A_i^{-1} R_i A u$, while $T_i u^* = R_i^T A_i^{-1} R_i f$.

We will use the operators $\boldsymbol{T}_i$ to construct new operator equations, using polynomials in $\boldsymbol{T}_i$, that are inherently well conditioned. We can calculate with these operators since we already know how to compute $\boldsymbol{T}_i \boldsymbol{u}^*$ without knowing $\boldsymbol{u}^*$, as well as $\boldsymbol{T}_i \boldsymbol{u}$ for any given $\boldsymbol{u}$. From these operator equations various preconditioners may be derived. Let $\mathcal{P}(\boldsymbol{T}_0, \boldsymbol{T}_1, \ldots, \boldsymbol{T}_p)$ be any polynomial in $\boldsymbol{T}_i$ such that $\mathcal{P}(0, 0, \ldots, 0) = 0$, then the action of that polynomial on the vector $\boldsymbol{u}^*$, $\mathcal{P}(\boldsymbol{T}_0, \boldsymbol{T}_1, \ldots, \boldsymbol{T}_p)\boldsymbol{u}^*$ can be calculated without knowing $\boldsymbol{u}^*$, similarly $\mathcal{P}(\boldsymbol{T}_0, \boldsymbol{T}_1, \ldots, \boldsymbol{T}_p)\boldsymbol{u}$ may be calculated for any given $\boldsymbol{u}$.

Let $\boldsymbol{g} = \mathcal{P}(\boldsymbol{T}_0, \boldsymbol{T}_1, \ldots, \boldsymbol{T}_p)\boldsymbol{u}^*$. The general abstract Schwarz method replaces (5.1) or (5.2) by the operator equation: find $\boldsymbol{u}$ such that

$$\mathcal{P}(\boldsymbol{T}_0, \boldsymbol{T}_1, \ldots, \boldsymbol{T}_p)\boldsymbol{u} = \boldsymbol{g},$$

which can be solved by, for instance, a Krylov subspace method. The reason for this construction is that we replace an ill-conditioned operator $\boldsymbol{A}$ with one that is much better conditioned. If we interpret the transformed operator $\mathcal{P}(\boldsymbol{T}_0, \boldsymbol{T}_1, \ldots, \boldsymbol{T}_p)$ as a preconditioned form of the operator $\boldsymbol{A}$, then we may write $\mathcal{P}(\boldsymbol{T}_0, \boldsymbol{T}_1, \ldots, \boldsymbol{T}_p) = \boldsymbol{B}\boldsymbol{A}$, where $\boldsymbol{B}$, the preconditioner, is defined implicitly. In matrix notation the application of B to a vector is always straightforward though explicit formulas for B are often confusing.

There are a wide variety of polynomials $\mathcal{P}(\cdot)$ which can be used. We consider four which seem to be of practical interest. Define $\boldsymbol{B}_i$ implicitly from the relationship $\boldsymbol{B}_i \boldsymbol{A} = \boldsymbol{T}_i$.

Example 1: Additive Schwarz Method

(see, e.g., Sections 1.1.2, 2.3, and 3.1)

$$\boldsymbol{B}\boldsymbol{A} = \mathcal{P}(\boldsymbol{T}_0, \boldsymbol{T}_1, \ldots, \boldsymbol{T}_p) = \boldsymbol{T}_0 + \boldsymbol{T}_1 + \cdots + \boldsymbol{T}_p.$$

The action of $\boldsymbol{B}$ on a function $\boldsymbol{r}$ may be calculated with

$$\boldsymbol{v} \leftarrow \sum_{i=0}^{p} \boldsymbol{B}_i \boldsymbol{r}.$$

$$\boldsymbol{B} = \sum_{i=0}^{p} \boldsymbol{B}_i.$$

Example 2: Multiplicative Schwarz Method

(see, e.g., Sections 1.1, 2.3, and 3.2).

$$BA = \mathcal{P}(T_0, T_1, \dots, T_p) = I - (I - T_p) \cdots (I - T_0).$$

The action of B on a function r may be calculated with

$$\begin{aligned} v &\leftarrow B_0 r \\ v &\leftarrow v + B_1(r - Av) \\ &\dots \\ v &\leftarrow v + B_p(r - Av). \end{aligned}$$

$$\begin{aligned} B &= (I - (I - B_p A) \cdots (I - B_0 A))A^{-1} \\ &= A^{-1}(I - (I - AB_p) \cdots (I - AB_0)). \end{aligned}$$

Example 3: Hybrid I Method

(see, e.g., Section 2.3)

$$BA = \mathcal{P}(T_0, T_1, \dots, T_p) = T_0 + I - (I - T_p) \cdots (I - T_1).$$

The action of B on a function r may be calculated with

$$\begin{aligned} v &\leftarrow B_1 r \\ v &\leftarrow v + B_2(r - Av) \\ &\dots \\ v &\leftarrow v + B_p(r - Av) \\ v &\leftarrow B_0 r + v. \end{aligned}$$

$$B = B_0 + (I - (I - B_p A) \cdots (I - B_1 A))A^{-1}.$$

Example 4: Hybrid II Method

(see, e.g., Section 2.3)

$$BA = \mathcal{P}(T_0, T_1, \dots, T_p) = T_0 + (I - T_0)(T_1 + \dots + T_p).$$

The action of B on a function r may be calculated with

$$\begin{aligned} v &\leftarrow \sum_{i=1}^{p} B_i r \\ v &\leftarrow v + B_0(r - Av). \end{aligned}$$

$$B = B_0 + (I - B_0 A)\left(\sum_{i=1}^{p} B_i\right).$$

For the multiplicative and hybrid methods, it is also possible to create symmetrized versions of the preconditioner. For instance, for the symmetric multiplicative Schwarz method the preconditioner is given by

$$\boldsymbol{B} = [I - (I - \boldsymbol{B}_0\boldsymbol{A}) \cdots (I - \boldsymbol{B}_p\boldsymbol{A})(I - \boldsymbol{B}_p\boldsymbol{A}) \cdots (I - \boldsymbol{B}_0\boldsymbol{A})]\boldsymbol{A}^{-1}.$$

It is possible to view the hybrid methods in a slightly different manner. They can be derived by first considering a subspace splitting into spaces V_0 and $V_* = V$. Then the subspace V_* is further split into subspaces V_1 to V_p. In this way hybrid I is simply additive Schwarz on V_0 and V_* with the bilinear form on V_* approximated by using a multiplicative Schwarz splitting into V_1 to V_p. Hybrid II is a multiplicative Schwarz method on V_0 and V_* where the bilinear form on V_* is approximated with an additive Schwarz splitting into V_1 to V_p.

A Schwarz algorithm is then defined by four components:

1. a set of (sub)spaces V_i,
2. a (possible) set of interpolation operators $\boldsymbol{I}_i : V_i \rightarrow V$,
3. a (possible) set of auxiliary bilinear forms $a_i(\cdot, \cdot)$, and
4. the order in which the (sub)space corrections are performed: that is, the choice of the polynomial $\mathcal{P}(\cdot)$.

Notes and References for Section 5.1

- The general concepts introduced in this chapter and their presentation in this abstract form are mostly due to Dryja and Widlund [Dry89], [Dry90] and their students and co-workers.
- See also Xu, for a slightly different presentation of this material [Xu92], and Griebel and Oswald [Gri95].
- Some of the ideas had been developed earlier by Lions [Lio78], [Lio88] and Nepomnyaschikh, see, for instance, [Nep84], [Nep86], and [Nep91a].
- Bramble [Bra93] has developed a similar framework, but devoting more attention to multigrid/multilevel algorithms with less emphasis on domain decomposition methods.
- The version hybrid I was introduced in X. Cai [Cai93].
- The version hybrid II was introduced in Mandel [Man91].

5.2 Abstract Convergence Analysis

In this section we introduce the tools that are used in the convergence analysis of Schwarz methods. For simplicity, we begin with symmetric positive definite problems: that is, $a(\cdot, \cdot)$ is symmetric and $a(\boldsymbol{u}, \boldsymbol{u}) \geq \alpha(\boldsymbol{u}, \boldsymbol{u})$, $\forall \boldsymbol{u} \in V$, for some $\alpha > 0$. In addition, assume that all of the $a_i(\cdot, \cdot)$ are symmetric and positive definite. We start with an elementary, but very powerful lemma. This lemma holds for both nested and nonnested spaces (in the case of nested subspaces we define $\boldsymbol{I}_i$ to be the trivial imbedding mapping).

Lemma 1: *Define* $T = \sum_{i=0}^{p} T_i$. *Then*

$$a(T^{-1}u, u) = \min_{\substack{u_i \in V_i \\ u = \sum_i I_i u_i}} \sum_i a_i(u_i, u_i). \tag{5.5}$$

PROOF: We first construct a particular decomposition of u which satisfies (5.5) exactly, and then prove

$$a(T^{-1}u, u) \le \sum_i a_i(u_i, u_i) \qquad \forall u_i \in V_i, \ \sum_i I_i u_i = u. \tag{5.6}$$

The proof only uses the definition of $\tilde{T}_i$ from (5.3), $T_i = I_i \tilde{T}_i$, and the Cauchy-Schwarz inequality.

Let $u_i = \tilde{T}_i T^{-1} u$; then $\sum_i I_i u_i = (\sum_i I_i \tilde{T}_i) T^{-1} u = u$ and

$$\begin{aligned}
\sum_i a_i(u_i, u_i) &= \sum_i a_i(\tilde{T}_i T^{-1}u, \tilde{T}_i T^{-1}u), \\
&= \sum_i a(T^{-1}u, T_i T^{-1}u), \\
&= a(T^{-1}u, u).
\end{aligned}$$

We now proceed to demonstrate (5.6),

$$\begin{aligned}
a(T^{-1}u, u) &= a(T^{-1}u, \sum_i I_i u_i), \\
&= \sum_i a(T^{-1}u, I_i u_i), \\
&= \sum_i a_i(\tilde{T}_i T^{-1}u, u_i), \\
&\le [\sum_i a_i(\tilde{T}_i T^{-1}u, \tilde{T}_i T^{-1}u)]^{1/2} [\sum_i a_i(u_i, u_i)]^{1/2}, \\
&= [\sum_i a(T^{-1}u, T_i T^{-1}u)]^{1/2} [\sum_i a_i(u_i, u_i)]^{1/2}, \\
&= a(T^{-1}u, u)^{1/2} [\sum_i a_i(u_i, u_i)]^{1/2}.
\end{aligned}$$

Therefore

$$a(T^{-1}u, u) \le \sum_i a_i(u_i, u_i) \qquad \forall u_i \in V_i, \ \sum_i I_i u_i = u.$$

□

The abstract convergence theory for symmetric problems centers around three parameters which measure the interactions of the subspaces V_i and the bilinear forms $a_i(\cdot, \cdot)$, and their suitability in the construction of preconditioners. We define the three parameters in the form of three assumptions.

Assumption 1: *Let C_0 be the minimum constant such that for all $\boldsymbol{u} \in V$ there exists a representation $\boldsymbol{u} = \sum_i \boldsymbol{I}_i \boldsymbol{u}_i$, $\boldsymbol{u}_i \in V_i$, with*

$$\sum_i a_i(\boldsymbol{u}_i, \boldsymbol{u}_i) \leq C_0^2 a(\boldsymbol{u}, \boldsymbol{u}).$$

If our **first parameter** C_0 can be bounded independently of the grid parameters (size and number of elements and subdomains), then the V_i are said to provide a **stable splitting** of V. We will see below that a C_0 near 1 is desirable. The quantities $a(\boldsymbol{u}, \boldsymbol{u})$ and $a_i(\boldsymbol{u}_i, \boldsymbol{u}_i)$ are referred to as the energies of $\boldsymbol{u}$ and $\boldsymbol{u}_i$. We see that this assumption together with Lemma 1 immediately provides C_0^{-2} as a lower bound on the spectrum of $\boldsymbol{T} = \sum_{i=0}^{p} T_i$.

Assumption 2: *Define $0 \leq \mathcal{E}_{ij} \leq 1$ to be the minimal values that satisfy*

$$|a(\boldsymbol{I}_i \boldsymbol{u}_i, \boldsymbol{I}_j \boldsymbol{u}_j)| \leq \mathcal{E}_{ij} a(\boldsymbol{I}_i \boldsymbol{u}_i, \boldsymbol{I}_i \boldsymbol{u}_i)^{1/2} a(\boldsymbol{I}_j \boldsymbol{u}_j, \boldsymbol{I}_j \boldsymbol{u}_j)^{1/2}$$
$$\forall \boldsymbol{u}_i \in V_i,\ \boldsymbol{u}_j \in V_j,\ i, j = 1, \ldots, p.$$

Define $\rho(\mathcal{E})$ to be the spectral radius of $\mathcal{E}$. Note that we do not include the subspace V_0.

The **second parameter** $\rho(\mathcal{E})$ is in some sense a measure of the orthogonality of the subspaces. The orthogonality of the subspaces V_i is measured by using the **strengthened Cauchy-Schwarz** inequality. When $\mathcal{E}_{ij}$ is zero, this implies that the subspaces V_i and V_j are orthogonal while $\mathcal{E}_{ij} = 1$ is the usual Cauchy-Schwarz inequality. Below, we will observe that a $\rho(\mathcal{E})$ near 1 is desirable. Also note that we have the inequality $\sum_{i,j} \mathcal{E}_{ij} x_i x_j \leq \rho(\mathcal{E}) \sum_i x_i^2$ since $\mathcal{E}_{ij} = \mathcal{E}_{ji}$ and $x^T \mathcal{E} x / x^T x \leq \rho(\mathcal{E})$. We will make use of this and similar inequalities several times.

Assumption 3: *Let $\omega \in [1, 2)$ be the minimum constant such that*

$$a(\boldsymbol{I}_i \boldsymbol{u}, \boldsymbol{I}_i \boldsymbol{u}) \leq \omega a_i(\boldsymbol{u}, \boldsymbol{u}) \qquad \forall \boldsymbol{u} \in V_i,\ i = 0, \ldots, p,$$

where we assume that the $a_i(\cdot, \cdot)$ are suitably scaled.

The **third parameter** ω is a one-sided measure of the approximation properties of the $a_i(\cdot, \cdot)$. In particular, if $a_i(\cdot, \cdot) = a(\cdot, \cdot)$ corresponding to the use of exact subdomain solvers, we can take $\omega = 1$. The assumed scaling of the bilinear forms $a_i(\cdot, \cdot)$ in order to restrict the values of ω means that we are not completely free to scale the $a_i(\cdot, \cdot)$ in order to decrease C_0. Note that C_0 includes a measure of the other side of the approximation properties of $a_i(\cdot, \cdot)$, see Assumption 1.

One choice for $a_i(\boldsymbol{u}, \boldsymbol{u})$ is simply

$$a_i(\boldsymbol{u}, \boldsymbol{u}) = \frac{1}{\omega} a(\boldsymbol{I}_i \boldsymbol{u}, \boldsymbol{I}_i \boldsymbol{u}).$$

In this case ω becomes a parameter which may be tuned to improve the convergence, much like the parameter in SOR. As with SOR there is no good general procedure for determining the best ω.

For a linear operator $\boldsymbol{L}$, which is self-adjoint with respect to $a(\cdot,\cdot)$, we use the Rayleigh quotient characterization of the extreme eigenvalues,

$$\lambda^A_{\min}(\boldsymbol{L}) = \min_{u\neq 0} \frac{a(\boldsymbol{Lu},\boldsymbol{u})}{a(\boldsymbol{u},\boldsymbol{u})}, \qquad \lambda^A_{\max}(\boldsymbol{L}) = \max_{u\neq 0} \frac{a(\boldsymbol{Lu},\boldsymbol{u})}{a(\boldsymbol{u},\boldsymbol{u})}.$$

The condition number of $\boldsymbol{L}$ is given by $\kappa(\boldsymbol{L}) = \lambda^A_{\max}(\boldsymbol{L})/\lambda^A_{\min}(\boldsymbol{L})$. Note that, for a given $a(\cdot,\cdot)$, the condition number of the operator is an intrinsic property of the operator and its relation to $a(\cdot,\cdot)$ and does not depend on the basis in which the operator is represented. Also note that, by definition,

$$\lambda^A_{\min}(\boldsymbol{L})a(\boldsymbol{u},\boldsymbol{u}) \leq a(\boldsymbol{Lu},\boldsymbol{u}) \leq \lambda^A_{\max}(\boldsymbol{L})a(\boldsymbol{u},\boldsymbol{u}) \qquad \forall \boldsymbol{u} \in V.$$

In order to apply the preconditioned conjugate gradient method for the operator $\boldsymbol{T}$, it must be symmetric with respect to $a(\cdot,\cdot)$, that is,

$$\boldsymbol{T}^T\boldsymbol{A} = \boldsymbol{A}\boldsymbol{T}.$$

Recall from Section 2.7.2 that even in the nonnested case the matrix representation of $\boldsymbol{T}_i$ is given by $T_i = R_i^T \tilde{A}_i^{-1} R_i A$. Thus it is easy to see that $T_i^T A = AT_i$. Hence $T^T A = AT$ where $T = \sum_i T_i$.

We verify this more formally for the abstract operators.

Lemma 2: *The operators $\boldsymbol{T}_i$, and hence $\boldsymbol{T} = \sum_i \boldsymbol{T}_i$, are self-adjoint in $a(\cdot,\cdot)$; that is,*

$$a(\boldsymbol{T}_i\boldsymbol{u},\boldsymbol{w}) = a(\boldsymbol{u},\boldsymbol{T}_i\boldsymbol{w}).$$

PROOF: This result follows immediately from the symmetry of $a(\cdot,\cdot)$ and the definition of $\boldsymbol{T}_i$.

$$\begin{aligned} a(\boldsymbol{T}_i\boldsymbol{u},\boldsymbol{w}) &= a(\boldsymbol{I}_i\tilde{\boldsymbol{T}}_i\boldsymbol{u},\boldsymbol{w}),\\ &= a(\boldsymbol{w},\boldsymbol{I}_i\tilde{\boldsymbol{T}}_i\boldsymbol{u}),\\ &= a_i(\tilde{\boldsymbol{T}}_i\boldsymbol{w},\tilde{\boldsymbol{T}}_i\boldsymbol{u}),\\ &= a_i(\tilde{\boldsymbol{T}}_i\boldsymbol{u},\tilde{\boldsymbol{T}}_i\boldsymbol{w}),\\ &= a(\boldsymbol{u},\boldsymbol{I}_i\tilde{\boldsymbol{T}}_i\boldsymbol{w}),\\ &= a(\boldsymbol{u},\boldsymbol{T}_i\boldsymbol{w}). \end{aligned}$$

□

We are now prepared to provide a bound on the condition number for the abstract additive Schwarz method in terms of the parameters in Assumptions 1, 2, and 3.

Lemma 3: *The abstract additive Schwarz method satisfies*

$$\kappa(\boldsymbol{BA}) \leq \omega[1+\rho(\mathcal{E})]C_0^2.$$

In particular, $1/C_0^2$ is a sharp lower bound on the smallest eigenvalue of $\boldsymbol{BA} = \sum_i \boldsymbol{T}_i$ and $\omega[1+\rho(\mathcal{E})]$ is a bound on the largest eigenvalue.

PROOF: The bound on the smallest eigenvalue follows immediately from Lemma 1 and Assumption 1.

We now bound the largest eigenvalue. We first treat the "standard" subspaces, $i = 1, \ldots, p$. The special subspace V_0 is treated separately. Observe that

$$\begin{aligned}
a(\sum_{i=1}^{p} T_i v, \sum_{i=1}^{p} T_i v) &= \sum_{i,j=1}^{p} a(T_i v, T_j v), \\
&\leq \sum_{i,j=1}^{p} \mathcal{E}_{ij} a(T_i v, T_i v)^{1/2} a(T_j v, T_j v)^{1/2}, \\
&\leq \rho(\mathcal{E}) \sum_{i=1}^{p} a(T_i v, T_i v), \\
&\leq \omega\rho(\mathcal{E}) \sum_{i=1}^{p} a_i(\tilde{T}_i v, \tilde{T}_i v), \\
&= \omega\rho(\mathcal{E}) \sum_{i=1}^{p} a(v, T_i v), \\
&= \omega\rho(\mathcal{E}) a(v, \sum_{i=1}^{p} T_i v), \\
&\leq \omega\rho(\mathcal{E}) a(v, v)^{1/2} a(\sum_{i=1}^{p} T_i v, \sum_{j=1}^{p} T_j v)^{1/2}.
\end{aligned}$$

Therefore

$$\begin{aligned}
a([\sum_{i=1}^{p} T_i]^2 v, v) &= a(\sum_{i=1}^{p} T_i v, \sum_{j=1}^{p} T_j v), \\
&\leq \omega^2 \rho^2(\mathcal{E}) a(v, v).
\end{aligned}$$

Thus the largest eigenvalue of $(\sum_{i=1}^{p} T_i)^2$ is bounded by $\omega^2\rho^2(\mathcal{E})$. Hence the largest eigenvalue of $\sum_{i=1}^{p} T_i$ is bounded by $\omega\rho(\mathcal{E})$ and

$$a(\sum_{i=1}^{p} T_i v, v) \leq \omega\rho(\mathcal{E}) a(v, v). \tag{5.7}$$

To estimate the largest eigenvalue of T_0 we use

$$\begin{aligned}
a(T_0 v, T_0 v) &\leq \omega a_0(\tilde{T}_0 v, \tilde{T}_0 v), \\
&= \omega a(v, T_0 v), \\
&\leq \omega a(v, v)^{1/2} a(T_0 v, T_0 v)^{1/2}.
\end{aligned} \tag{5.8}$$

Thus

$$a(T_0 v, T_0 v) \leq \omega^2 a(v, v),$$

which implies, again, as above, that

$$a(\boldsymbol{T}_0\boldsymbol{v}, \boldsymbol{v}) \leq \omega a(\boldsymbol{v}, \boldsymbol{v});$$

that is, the largest eigenvalue of $\boldsymbol{T}_0$ is bounded by ω. Combine this with (5.7) to complete the proof for the upper bound. □

In the case of nested subspaces, $V_i \subset V$, it is often convenient to view the subspaces as the range of an interpolation operator. The combination of Lemma 1 and Assumption 1 can then be viewed as a statement saying that this interpolation should not increase the energy of any function by more than a small amount, unless that function is also contained in another of the subspaces which treats it well.

Lemma 4: *The symmetric abstract multiplicative Schwarz method satisfies*

$$\kappa(\boldsymbol{BA}) \leq \frac{[1 + 2\omega^2\rho^2(\mathcal{E})]C_0^2}{2-\omega}.$$

PROOF: Let $\boldsymbol{E}_{-1} = I$ and let

$$\boldsymbol{E}_i = (I - \boldsymbol{T}_i)\boldsymbol{E}_{i-1} = (I - \boldsymbol{T}_i)\cdots(I - \boldsymbol{T}_0).$$

Then $\boldsymbol{BA} = I - \boldsymbol{E}_p^T\boldsymbol{E}_p$ where the transpose is in the $a(\cdot,\cdot)$ inner product; that is, $a(\boldsymbol{E}_p^T\boldsymbol{w}, \boldsymbol{v}) = a(\boldsymbol{w}, \boldsymbol{E}_p\boldsymbol{v})$. We first demonstrate that $\lambda_{\max}^A(\boldsymbol{BA}) \leq 1$.

$$\begin{aligned}\lambda_{\max}^A(\boldsymbol{BA}) &= \max_{u\neq 0}\frac{a(\boldsymbol{BAu}, \boldsymbol{u})}{a(\boldsymbol{u}, \boldsymbol{u})},\\ &= \max_{u\neq 0}\frac{a((I - \boldsymbol{E}_p^T\boldsymbol{E}_p)\boldsymbol{u}, \boldsymbol{u})}{a(\boldsymbol{u}, \boldsymbol{u})},\\ &= 1 - \min_{u\neq 0}\frac{a(\boldsymbol{E}_p\boldsymbol{u}, \boldsymbol{E}_p\boldsymbol{u})}{a(\boldsymbol{u}, \boldsymbol{u})},\\ &\leq 1.\end{aligned}$$

This follows from the definition of $\boldsymbol{BA}$, Lemma 2, and the fact that $a(\cdot,\cdot)$ is positive definite; hence the second term is positive.

The proof for the lower bound consists of algebraic manipulations, results from the previous proof, and applications of the Cauchy-Schwarz inequality.

As above

$$\begin{aligned}a(\boldsymbol{BAv}, \boldsymbol{v}) &= a(\boldsymbol{v}, \boldsymbol{v}) - a(\boldsymbol{E}_p\boldsymbol{v}, \boldsymbol{E}_p\boldsymbol{v}),\\ &= \sum_{i=0}^{p}[a(\boldsymbol{E}_{i-1}\boldsymbol{v}, \boldsymbol{E}_{i-1}\boldsymbol{v}) - a(\boldsymbol{E}_i\boldsymbol{v}, \boldsymbol{E}_i\boldsymbol{v})],\\ &= \sum_{i=0}^{p}[a(\boldsymbol{E}_{i-1}\boldsymbol{v}, \boldsymbol{E}_{i-1}\boldsymbol{v}) - a((\boldsymbol{E}_{i-1} - \boldsymbol{T}_i\boldsymbol{E}_{i-1})\boldsymbol{v}, (\boldsymbol{E}_{i-1} - \boldsymbol{T}_i\boldsymbol{E}_{i-1})\boldsymbol{v})],\end{aligned}$$

$$= \sum_{i=0}^{p} [2a(\boldsymbol{E}_{i-1}\boldsymbol{v}, \boldsymbol{T}_i\boldsymbol{E}_{i-1}\boldsymbol{v}) - a(\boldsymbol{T}_i\boldsymbol{E}_{i-1}\boldsymbol{v}, \boldsymbol{T}_i\boldsymbol{E}_{i-1}\boldsymbol{v})],$$

$$\geq (2-\omega) \sum_{i=0}^{p} a(\boldsymbol{E}_{i-1}\boldsymbol{v}, \boldsymbol{T}_i\boldsymbol{E}_{i-1}\boldsymbol{v}). \tag{5.9}$$

In the last line we have used the inequality

$$\begin{aligned} a(\boldsymbol{T}_i\boldsymbol{E}_{i-1}\boldsymbol{v}, \boldsymbol{T}_i\boldsymbol{E}_{i-1}\boldsymbol{v}) &\leq \omega a_i(\tilde{\boldsymbol{T}}_i\boldsymbol{E}_{i-1}\boldsymbol{v}, \tilde{\boldsymbol{T}}_i\boldsymbol{E}_{i-1}\boldsymbol{v}), \\ &= \omega a(\boldsymbol{E}_{i-1}\boldsymbol{v}, \boldsymbol{T}_i\boldsymbol{E}_{i-1}\boldsymbol{v}), \end{aligned} \tag{5.10}$$

which follows from Assumption 3.

By using the definition of $\boldsymbol{E}_i$, we note the equality

$$I = \boldsymbol{T}_0 + \boldsymbol{E}_{i-1} + \sum_{j=1}^{i-1} \boldsymbol{T}_j\boldsymbol{E}_{j-1}.$$

Using this, the Cauchy-Schwarz inequality and Assumptions 2 and 3,

$$\begin{aligned} a(\boldsymbol{T}_i\boldsymbol{v}, \boldsymbol{v}) &= a(\boldsymbol{T}_i\boldsymbol{v}, \boldsymbol{T}_0\boldsymbol{v}) + a(\boldsymbol{T}_i\boldsymbol{v}, \boldsymbol{E}_{i-1}\boldsymbol{v}) + \sum_{j=1}^{i-1} a(\boldsymbol{T}_i\boldsymbol{v}, \boldsymbol{T}_j\boldsymbol{E}_{j-1}\boldsymbol{v}), \\ &\leq a(\boldsymbol{v}, \boldsymbol{T}_i\boldsymbol{T}_0\boldsymbol{v}) + a(\boldsymbol{v}, \boldsymbol{T}_i\boldsymbol{E}_{i-1}\boldsymbol{v}) \\ &\quad + \sum_{j=1}^{i-1} \mathcal{E}_{ij} a(\boldsymbol{T}_i\boldsymbol{v}, \boldsymbol{T}_i\boldsymbol{v})^{1/2} a(\boldsymbol{T}_j\boldsymbol{E}_{j-1}\boldsymbol{v}, \boldsymbol{T}_j\boldsymbol{E}_{j-1}\boldsymbol{v})^{1/2}, \\ &\leq a(\boldsymbol{v}, \boldsymbol{T}_i\boldsymbol{T}_0\boldsymbol{v}) + a(\boldsymbol{v}, \boldsymbol{T}_i\boldsymbol{E}_{i-1}\boldsymbol{v}) \\ &\quad + \omega \sum_{j=1}^{i-1} \mathcal{E}_{ij} a_i(\tilde{\boldsymbol{T}}_i\boldsymbol{v}, \tilde{\boldsymbol{T}}_i\boldsymbol{v})^{1/2} a_j(\tilde{\boldsymbol{T}}_j\boldsymbol{E}_{j-1}\boldsymbol{v}, \tilde{\boldsymbol{T}}_j\boldsymbol{E}_{j-1}\boldsymbol{v})^{1/2}, \\ &\leq a_i(\tilde{\boldsymbol{T}}_i\boldsymbol{v}, \tilde{\boldsymbol{T}}_i\boldsymbol{v})^{1/2} [a_i(\tilde{\boldsymbol{T}}_i\boldsymbol{T}_0\boldsymbol{v}, \tilde{\boldsymbol{T}}_i\boldsymbol{T}_0\boldsymbol{v})^{1/2} + a_i(\tilde{\boldsymbol{T}}_i\boldsymbol{E}_{i-1}\boldsymbol{v}, \tilde{\boldsymbol{T}}_i\boldsymbol{E}_{i-1}\boldsymbol{v})^{1/2} \\ &\quad + \omega \sum_{j=1}^{i-1} \mathcal{E}_{ij} a_i(\tilde{\boldsymbol{T}}_j\boldsymbol{E}_{j-1}\boldsymbol{v}, \tilde{\boldsymbol{T}}_j\boldsymbol{E}_{j-1}\boldsymbol{v})^{1/2}], \\ &\leq a(\boldsymbol{T}_i\boldsymbol{v}, \boldsymbol{v})^{1/2} [a(\boldsymbol{T}_i\boldsymbol{T}_0\boldsymbol{v}, \boldsymbol{T}_0\boldsymbol{v})^{1/2} + a(\boldsymbol{T}_i\boldsymbol{E}_{i-1}\boldsymbol{v}, \boldsymbol{E}_{i-1}\boldsymbol{v})^{1/2} \\ &\quad + \omega \sum_{j=1}^{i-1} \mathcal{E}_{ij} a(\boldsymbol{T}_j\boldsymbol{E}_{j-1}\boldsymbol{v}, \boldsymbol{E}_{j-1}\boldsymbol{v})^{1/2}], \\ &\leq a(\boldsymbol{T}_i\boldsymbol{v}, \boldsymbol{v})^{1/2} [a(\boldsymbol{T}_i\boldsymbol{T}_0\boldsymbol{v}, \boldsymbol{T}_0\boldsymbol{v})^{1/2} + \omega \sum_{j=1}^{i} \mathcal{E}_{ij} a(\boldsymbol{T}_j\boldsymbol{E}_{j-1}\boldsymbol{v}, \boldsymbol{E}_{j-1}\boldsymbol{v})^{1/2}]. \end{aligned}$$

Now cancel the common factor and square both sides and use the relation $(a+b)^2 \leq 2a^2 + 2b^2$ to obtain

$$a(\boldsymbol{T}_i\boldsymbol{v}, \boldsymbol{v}) \leq 2a(\boldsymbol{T}_i\boldsymbol{T}_0\boldsymbol{v}, \boldsymbol{T}_0\boldsymbol{v}) + 2\omega^2 [\sum_{j=1}^{i} \mathcal{E}_{ij} a(\boldsymbol{T}_j\boldsymbol{E}_{j-1}\boldsymbol{v}, \boldsymbol{E}_{j-1}\boldsymbol{v})^{1/2}]^2.$$

Next, sum over i; extending the previous sum to p will only increase the right hand side,

$$\sum_{i=1}^{p} a(\boldsymbol{T}_i\boldsymbol{v}, \boldsymbol{v}) \leq 2\sum_{i=1}^{p} a(\boldsymbol{T}_i\boldsymbol{T}_0\boldsymbol{v}, \boldsymbol{T}_0\boldsymbol{v}) + 2\omega^2 \sum_{i=1}^{p}[\sum_{j=1}^{p} \mathcal{E}_{ij} a(\boldsymbol{T}_j\boldsymbol{E}_{j-1}\boldsymbol{v}, \boldsymbol{E}_{j-1}\boldsymbol{v})^{1/2}]^2.$$

For the first term we use the bound (5.7) combined with inequality (5.8) to obtain

$$2a([\sum_{i=1}^{p} \boldsymbol{T}_i]\boldsymbol{T}_0\boldsymbol{v}, \boldsymbol{T}_0\boldsymbol{v}) \leq 2\omega^2\rho(\mathcal{E})a(\boldsymbol{T}_0\boldsymbol{v}, \boldsymbol{v}).$$

The second term is of the form $(\mathcal{E}x)^T(\mathcal{E}x) \leq \rho^2(\mathcal{E})x^T x$ for a vector x with components $a(\boldsymbol{T}_j\boldsymbol{E}_{j-1}\boldsymbol{v}, \boldsymbol{E}_{j-1}\boldsymbol{v})^{1/2}$. We now add the term $a(\boldsymbol{T}_0\boldsymbol{v}, \boldsymbol{v})$ to both sides, observing that $a(\boldsymbol{T}_0\boldsymbol{v}, \boldsymbol{v}) = a(\boldsymbol{T}_0\boldsymbol{E}_{-1}\boldsymbol{v}, \boldsymbol{E}_{-1}\boldsymbol{v})$, giving

$$\sum_{i=0}^{p} a(\boldsymbol{T}_i\boldsymbol{v}, \boldsymbol{v}) \leq [1 + 2\omega^2\rho^2(\mathcal{E})] \sum_{i=0}^{p} a(\boldsymbol{T}_i\boldsymbol{E}_{i-1}\boldsymbol{v}, \boldsymbol{E}_{i-1}\boldsymbol{v}).$$

Here we assume, without loss of generality, that $\rho(\mathcal{E}) \geq 1$ and hence $\rho^2(\mathcal{E}) \geq \rho(\mathcal{E})$. Finally, we use the lower bound from the additive case given in Lemma 3, together with the inequality (5.9), to obtain

$$\frac{2-\omega}{[1 + 2\omega^2\rho^2(\mathcal{E})]C_0^2} a(\boldsymbol{v}, \boldsymbol{v}) \leq a(\boldsymbol{BAv}, \boldsymbol{v}).$$

Since the upper bound was 1, the desired result is proved. □

It is also possible to derive bounds for the hybrid methods. We present one example below.

Lemma 5: *When the $\boldsymbol{T}_i$ are projections, then the condition number of **A**, preconditioned with the symmetrized hybrid II Schwarz method, is less than or equal to that of **A** preconditioned by the additive Schwarz preconditioner.*

This lemma confirms a very intuitive situation: as we include more multiplicative features into our algorithms, we can expect the condition number of the preconditioned iteration operator ***BA*** to decrease. The two extreme cases are a completely parallel, purely additive method with the slowest convergence and a completely multiplicative algorithm, with the fastest convergence.

Notes and References for Section 5.2

- The importance of Assumption 1 was first noticed by Matsokin and Nepomnyaschikh [Mats85] and Lions [Lio88].
- The convergence theory for multiplicative methods was developed in Bramble, Pasciak, Wang, and Xu [Bra91b]. Earlier partial results were obtained by Mathew [Math89].
- The proofs for Lemmas 3 and 4 are essentially those given in Dryja and Widlund [Dry95].

- The proof for Lemma 5 is given in Mandel [Man91].
- The complete relationship between the multiplicative and additive Schwarz methods with two subspaces has been completely understood for several years; see, e.g., Bjørstad [Bjø89] and Bjørstad and Mandel [Bjø91].
- Lemma 1 appears to have been first written in that form by Zhang [Zha91], though he apparently only considered the case when the T_i are projection operators and the spaces are nested. The general formula appears, though not explicitly, in Griebel and Oswald [Gri95].
- The strengthened Cauchy-Schwarz inequality is often used in analyzing multilevel methods; see, for example, Eijkhout and Vassilevski [Eij91].
- A survey of some of the material presented here has appeared in Xu [Xu92]. This was also the first paper to include the analysis of the multiplicative algorithms in the same framework as the additive methods.
- The paper of Griebel and Oswald [Gri95] contains a more delicate analysis of the multiplicative version of the algorithms, using the techniques introduced in Section 3.5 with sharper results than those given here.

5.3 Analysis of Standard Methods

In this section we give convergence proofs for several standard domain decomposition and multilevel methods using the abstract theory introduced above. The technical details are often a bit overwhelming; we therefore supplement the proofs with a more informal discussion.

All of the convergence proofs are organized in the same manner. We first construct a partition of $\boldsymbol{u}$ into the subspaces and next bound the energy of these pieces. This gives us a bound for Assumption 1. To verify Assumption 2 we determine bounds on the angles between the subspaces. Finally we determine the one-sided bound on $a_i(\cdot,\cdot)$ needed in Assumption 3.

For simplicity in all the convergence proofs, except for the multilevel diagonal scaling method, we consider only the following configuration of subdomains. A domain $\Omega \in R^2$ (or R^3) is the union of shape regular triangular (tetrahedral) superelements of diameter $O(H)$. Assume that the superelements are further triangulated into shape regular elements of diameter $O(h)$. In addition, we assume that piecewise linear finite elements are used in the discretization. Many extensions of these results are possible. At the end of each section we try to indicate references to where they have been carried out.

5.3.1 Overlapping Schwarz Methods

We first consider the simple, second order Poisson's equation,

$$\begin{aligned} -\Delta \boldsymbol{u} &= f \quad \text{in } \Omega, \\ \boldsymbol{u} &= 0 \quad \text{on } \partial\Omega. \end{aligned}$$

To put this problem in the form needed for the convergence theory we multiply by a test function $v \in H_0^1(\Omega)$, integrate and apply a Green's formula to obtain

$$a(\boldsymbol{u}, v) = \int_\Omega (\nabla \boldsymbol{u}) \cdot (\nabla v)\, dx = \int_\Omega f v\, dx = (f, v) \qquad \forall v \in H_0^1(\Omega).$$

Let $V^h \subset H_0^1(\Omega)$ be the finite element space defined by piecewise linear elements on a shape regular, quasi-uniform triangulation. Then our finite dimensional variational problem is given by: Find $\boldsymbol{u}^h \in V^h$ such that

$$a(\boldsymbol{u}^h, v^h) = \int_\Omega (\nabla \boldsymbol{u}^h) \cdot (\nabla v^h)\, dx = \int_\Omega f v^h\, dx \qquad \forall v^h \in V^h.$$

By construction, the $a(\cdot, \cdot)$ is symmetric and positive definite, so the abstract algorithms and theory in the previous section may be applied. Let us briefly recall the standard notation for norms and seminorms that will be used in this chapter. With reference to the model problem above, we write

$$\begin{aligned}
||\boldsymbol{u}||_{L^2(\Omega)} &= \left(\int_\Omega \boldsymbol{u}^2 dx \right)^{1/2}, \\
||\boldsymbol{u}||_{L^\infty(\Omega)} &= \sup_{x \in \Omega} |\boldsymbol{u}(x)|, \\
||\boldsymbol{u}||_{H^1(\Omega)} &= [||\boldsymbol{u}||^2_{L^2(\Omega)} + ||\nabla \boldsymbol{u}||^2_{L^2(\Omega)}]^{1/2}, \\
|\boldsymbol{u}|_{H^1(\Omega)} &= ||\nabla \boldsymbol{u}||_{L^2(\Omega)}, \\
||\boldsymbol{u}||_a &= a(\boldsymbol{u}, \boldsymbol{u})^{1/2}.
\end{aligned}$$

Here the last norm is generated by a possibly more general bilinear form $a(\cdot, \cdot)$ than our model problem above. A reader unfamiliar with these concepts should consult a standard reference on the finite element method.

The subspaces used in the theory (and the algorithm) for the two level overlapping Schwarz method are the coarse grid subspace V^H and the local subspaces associated with the overlapping regions, $V_i^h = V^h \cap H_0^1(\Omega_i)$; see Sections 1.6 and 2.8.

The following theorem shows that for generous overlap between the subdomains the condition number of the preconditioned problem using the two level overlapping Schwarz method does not grow as the mesh is refined. This means that the maximum number of iterations of the preconditioned conjugate gradient method needed to achieve a fixed tolerance does not grow as the mesh is refined. However, since the truncation error will decrease as one refines the mesh, the number of iterations required to reach the truncation error may increase. See Section 3.3, where this is dealt with through the use of nested iteration.

Throughout the rest of this chapter we will drop the superscript h from our finite element functions. That is, $\boldsymbol{u}$ may (possibly) denote a finite element function.

Theorem 1: *For the two level overlapping Schwarz method (Algorithms 2.3.2 and 2.3.3), with exact subdomain solves, if the overlap is uniformly of width $O(H)$ then the condition number of the preconditioned system is bounded independently of h and H.*

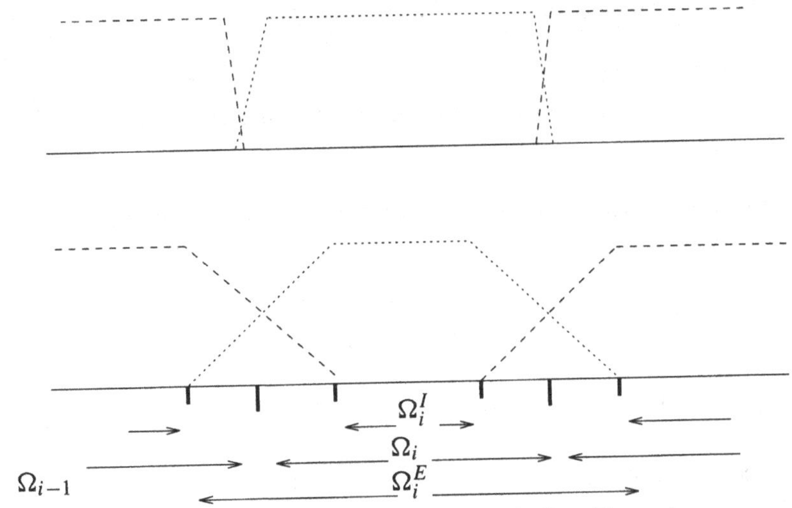

Figure 5.1. Sample Decompositions in One Dimension

We first construct an explicit decomposition of all $u \in V^h$ and verify Assumption 1 with that decomposition. We will discuss Assumption 2 below. Assumption 3 follows immediately with $\omega = 1$ because $a_i(\cdot, \cdot) = a(\cdot, \cdot)$ (we include the theorem and proof only for the case of nested subspaces).

More generally, if we replace the local and coarse grid solvers with spectrally equivalent forms the same uniform bound on the convergence rate is retained.

Let Ω_i denote the nonoverlapping subdomains and Ω_i^E the extended, overlapping subdomains. We now indicate several possible procedures to obtain a decomposition of u. All, except the last, can produce functions that have much higher energy than the initial function and thus are unacceptable for the proof.

(**D1**) One might first consider the simplest partition

$$\tilde{u}_i(x) = u(x) \quad \text{if } x \in \Omega_i,$$
$$\tilde{u}_i(x) = 0 \quad \text{if } x \in \Omega \setminus \Omega_i.$$

These functions are, however, not finite element functions; that is, they do not live in V^h, since they are not continuous. Therefore they do not provide a valid decomposition of u.

(**D2**) We can modify the decomposition introduced above by interpolating the functions back onto V^h. Thus we could define

$$u_i = I^h \tilde{u}_i.$$

Then $u = I^h \sum_i \tilde{u}_i = \sum_i I^h \tilde{u}_i = \sum_i u_i$ and $u_i \in V^h$. Unfortunately, these functions have generally steep gradients near the boundaries of each subdomain (see Figure 5.1); thus they will still have larger energy than u. Hence the bound in Assumption 1 will be poor: that is, it will depend on the mesh spacing.

(**D3**) To reduce the gradients near the subdomain boundaries we take advantage of the generous overlap between the subdomains. Let Ω_i^I denote the interior portion of

Ω_i that is not covered by any other Ω_i^E. We could define

$$\begin{aligned} u_i(x) &= u(x) && \text{if } x \in \Omega_i^I, \\ u_i(x) &= \text{smoothly decreasing toward } \partial\Omega_i^E && \text{if } x \in \Omega_i^E \setminus \Omega_i^I, \\ u_i(x) &= 0 && \text{if } x \in \Omega \setminus \Omega_i^E. \end{aligned}$$

To make this construction more mathematically rigorous we introduce a smooth partition of unity: that is, a set of smooth functions θ_i on Ω such that $\sum \theta_i = 1$,

$$\begin{aligned} \theta_i(x) &= 1 && \text{for } x \in \Omega_i^I, \\ \theta_i(x) &= 0 && \text{for } x \in \Omega \setminus \Omega_i^E, \end{aligned}$$

and the gradient of θ_i is not too large, specifically $|\nabla\theta_i|^2_{L^\infty} \leq C/H^2$. Using this partition of unity we can define $u_i = I^h(\theta_i u)$.

We now need to estimate the energy of u_i. Unfortunately, it can still be rather large. Consider a function u that is constant and is equal to $1/\epsilon$ (for any small ϵ) over the domain Ω_i^E. The energy of u in Ω_i^E is zero but the energy of u_i (which is $O(\frac{H^d}{H^2}\frac{1}{\epsilon^2})$, in d dimensions, $d = 2, 3$) can be made arbitrarily large by decreasing ϵ. Thus we cannot bound the energy of u_i by the energy of u restricted to Ω_i^E, independent of ϵ. We can, however, bound it by the energy of u over the whole domain. This is not good enough, however, because if $a(u_i, u_i) \leq Ca(u, u)$ then $\sum_{i=1}^p a(u_i, u_i) \leq Cpa(u, u)$ and the bound in Assumption 1 will depend on the number of subdomains, p.

(D4) We need a way of extracting the troublesome constant functions before multiplying by the partition of unity functions. This is the role of the coarse grid space; see also the introduction to Chapter 2. We could try using $u_0 = I^H u$, the coarse grid interpolant of u, and then decompose the "leftover" by using the partition of unity, $u_i = I^h(\theta_i(u - u_0))$.

Alas, this is a bad idea since I^H can increase the energy of a function drastically. To see this, consider the function $u = \phi_j^h$ where ϕ_j^h is a fine grid finite element basis function associated with a coarse grid node x_k. Then $I^H u = \phi_k^H$, the coarse grid finite element basis function associated with the same node. Now, in three dimensions,

$$a(u, u) = \int_\Omega (\nabla\phi_j^h)^2\, dx = O(h),$$

while its interpolant has energy

$$a(I^H u, I^H u) = \int_\Omega (\nabla\phi_k^H)^2\, dx = O(H).$$

Thus, in three dimensions, the decomposition **(D4)** results in a bound on C_0^2 that is at best $O(H/h)$ for Assumption 1.

(D5) We need another linear operator $Q_H : V^h \to V^H$ that better extracts the "smooth" part of u. A natural choice for $Q_H u$ would be the element of V^H that is closest to u as measured in the L^2-norm. This is the L^2-projection of u (see Section 1.6.1).

$$Q_H u = \arg\min_{v \in V^H} ||u - v||^2_{L^2(\Omega)},$$

or, equivalently,

$$(Q_H \boldsymbol{u}, \boldsymbol{v})_{L^2(\Omega)} = (\boldsymbol{u}, \boldsymbol{v})_{L^2(\Omega)} \qquad \forall \boldsymbol{v} \in V^H.$$

Thus we propose the decomposition,

$$\begin{aligned} \boldsymbol{u}_0 &= Q_H \boldsymbol{u}, \\ \boldsymbol{u}_i &= I^h[\theta_i(\boldsymbol{u} - \boldsymbol{u}_0)]. \end{aligned}$$

This is the decomposition that is used in the proof. It is important to note that the L^2-projection is only needed for the proof; it is not used in the implementation of the algorithm.

For this decomposition to work, it is obvious (since we must be able to bound the energy of $\boldsymbol{u}_0$ in terms of $\boldsymbol{u}$) that we will need

$$|Q_H \boldsymbol{u}|_{H^1(\Omega)} \le C|\boldsymbol{u}|_{H^1(\Omega)}.$$

Fortunately, the H^1-stability of the L^2-projection is known. Proving this goes beyond the scope of this book; we refer the reader to Bramble and Xu [Bra91c] for a proof of this result. We will see below that for the convergence proof to work, $Q_H \boldsymbol{u}$ must also approximate $\boldsymbol{u}$. Formally,

$$||\boldsymbol{u} - Q_H \boldsymbol{u}||_{L^2(\Omega)} \le CH|\boldsymbol{u}|_{H^1(\Omega)}.$$

The L^2-projection operator works by essentially smoothing out the function $\boldsymbol{u}$ by averaging it over the coarse grid elements. Other local averaging operators are also possible, for instance the Clément interpolation [Cia78] (Section 3.2.3) or Strang smoothing [Str72]. A drawback of using the L^2-projection operator is that it requires the coarse grid to be quasi-uniform (all elements must be roughly the same size). The Clément interpolation merely requires that the elements be shape regular; it may be used on highly graded meshes.

The proof of Theorem 1 requires one more technical result, an **inverse inequality**. These inequalities are often used in finite element analysis for bounding stronger norms by weaker ones. For instance, the inverse inequality needed below states (see, e.g., Lemma 7.3 in Johnson [Joh87]) that for all piecewise linear finite element functions $\boldsymbol{u} \in V^h$ there exists a constant C that depends only on the element shape so that for each element K

$$|\boldsymbol{u}|^2_{H^1(K)} \le Ch^{-2}||\boldsymbol{u}||^2_{L^2(K)}.$$

This completes the discussion of how Assumption 1 is verified.

Assumption 2 is verified by observing that if we color the subdomains (see Section 1.3.1), then all of the projections $\boldsymbol{T}_i$ of a particular color are independent. That is, $a(\boldsymbol{T}_i\boldsymbol{u}, \boldsymbol{T}_j\boldsymbol{u}) = 0$, for all subdomains Ω_j of the same color as Ω_i. Thus the norm of $\sum_{i\in\text{color}} \boldsymbol{T}_i \le \omega$. So the norm of the sum of all the $\boldsymbol{T}_i$, $i > 0$, or equivalently $\omega\rho(\mathcal{E})$, is bounded from above by 1 plus the number of colors times ω. In this proof ω is exactly 1 since exact subdomain solvers are used. The extra 1 comes from the coarse grid space; see the proof below.

PROOF OF THEOREM 1: We now construct a bound on the smallest eigenvalue, $\lambda^A_{\min}(\boldsymbol{BA})$, of the preconditioned system for the two level overlapping additive Schwarz method; that is, we calculate C_0^2 in Assumption 1. Let $Q_H\boldsymbol{u}$ be the L^2-projection of $\boldsymbol{u}$ onto V^H. We know that Q_H satisfies

$$||\boldsymbol{u} - Q_H\boldsymbol{u}||^2_{L^2(\Omega)} \leq CH^2|\boldsymbol{u}|^2_{H^1(\Omega)} \tag{5.11}$$

and

$$|\boldsymbol{u} - Q_H\boldsymbol{u}|^2_{H^1(\Omega)} \leq C|\boldsymbol{u}|^2_{H^1(\Omega)}. \tag{5.12}$$

We then define

$$\boldsymbol{u}_0 = Q_H\boldsymbol{u}, \qquad \boldsymbol{w} = \boldsymbol{u} - \boldsymbol{u}_0$$

and

$$\boldsymbol{u}_i = I^h(\theta_i\boldsymbol{w}).$$

Here I^h is the linear interpolation operator onto the space V^h and the θ_i form a partition of unity with $\theta_i \in C_0^\infty(\Omega_i^E)$, $0 \leq \theta_i \leq 1$ and $\sum_{i=1}^p \theta_i = 1$. In fact, we need not require $\theta_i \in C_0^\infty(\Omega_i^E)$; having θ_i in the space of piecewise linear finite element functions is enough. Since I^h is a linear operator, it follows immediately that

$$\boldsymbol{u} = \sum_{i=0}^{p} \boldsymbol{u}_i.$$

Because of the generous overlap between subregions, we can ensure that the gradients of θ_i are well behaved. That is, θ_i can be constructed so that its gradients never grow faster than $|\nabla\theta_i|^2_{L^\infty} \leq C/H^2$. If we let K represent any single element in the triangulation this can be expressed as

$$||\theta_i - \bar{\theta}_i||^2_{L^\infty(K)} \leq C(h/H)^2. \tag{5.13}$$

Here $\bar{\theta}_i$ is the average of θ_i on element K.

We now estimate the H^1 norm of $\boldsymbol{u}_i$ over a single element.

$$\begin{aligned} |\boldsymbol{u}_i|^2_{H^1(K)} &= |I^h[\bar{\theta}_i\boldsymbol{w} + (\theta_i - \bar{\theta}_i)\boldsymbol{w}]|^2_{H^1(K)}, \\ &\leq 2|\bar{\theta}_i\boldsymbol{w}|^2_{H^1(K)} + 2|I^h[(\theta_i - \bar{\theta}_i)\boldsymbol{w}]|^2_{H^1(K)}, \end{aligned}$$

which can be bounded by using an inverse inequality by

$$|\boldsymbol{u}_i|^2_{H^1(K)} \leq 2|\bar{\theta}_i\boldsymbol{w}|^2_{H^1(K)} + Ch^{-2}||I^h[(\theta_i - \bar{\theta}_i)\boldsymbol{w}]||^2_{L^2(K)}.$$

Use (5.13) and the trivial inequality $||\bar{\theta}_i||_{L^\infty} \leq 1$ to obtain

$$|\boldsymbol{u}_i|^2_{H^1(K)} \leq 2|\boldsymbol{w}|^2_{H^1(K)} + CH^{-2}||\boldsymbol{w}||^2_{L^2(K)}.$$

Since a finite bounded number (independent of h and H) of $\boldsymbol{u}_i$ is nonzero for any element K, we obtain, when summing over i,

$$\sum_{i=1}^{p} |\boldsymbol{u}_i|^2_{H^1(K)} \leq C|\boldsymbol{w}|^2_{H^1(K)} + CH^{-2}||\boldsymbol{w}||^2_{L^2(K)}.$$

Next sum over the elements K,

$$\sum_{i=1}^{p} |\boldsymbol{u}_i|^2_{H^1(\Omega)} \leq C|\boldsymbol{w}|^2_{H^1(\Omega)} + CH^{-2}||\boldsymbol{w}||^2_{L^2(\Omega)}.$$

To finish the argument, we use (5.11) and (5.12) to obtain

$$\sum_{i=0}^{p} |\boldsymbol{u}_i|^2_{H^1(\Omega)} \leq C_0^2|\boldsymbol{u}|^2_{H^1(\Omega)}.$$

We have now verified that the first parameter C_0^2 is independent of h and H.

The second parameter can be estimated by using a coloring argument. We know that we can color the subdomains with N_c colors independently of h and H. Taking the coarse grid into account we therefore group the $\boldsymbol{T}_i$ into $N_c + 1$ classes so that all the $\boldsymbol{T}_i$ in a particular class are independent. Hence, $\rho(\mathcal{E}) \leq N_c + 1$. We have already noted that the third parameter $w = 1$. □

The reason we need to introduce the $\bar{\theta}_i$ and separate the estimate into two parts is that if we applied the inverse inequality directly to $|I^h(\theta_i \boldsymbol{w})|_{H^1(K)}$ we could not cancel the h^{-2} term and the resulting bound would be a disappointing H^2/h^2.

Numerical studies indicate reasonably good convergence even with small overlap. In order to obtain sharper theoretical results for the overlapping Schwarz methods we first introduce a useful lemma.

Lemma 6: *Let Ω_i be a shape regular region, in R^2 or R^3, of diameter H and let $\Gamma_{\delta,i}$ be a strip along its boundary of width $\delta > 0$ (see Figure 5.2). Then*

$$||\boldsymbol{u}||^2_{L^2(\Gamma_{\delta,i})} \leq C\delta^2 \left[\left(1 + \frac{H}{\delta}\right) |\boldsymbol{u}|^2_{H^1(\Omega_i)} + \frac{1}{H\delta}||\boldsymbol{u}||^2_{L^2(\Omega_i)} \right].$$

The proof is essentially a modification of the elementary proofs of Poincaré type inequalities.

Using this result we can modify Theorem 1 and its proof.

Theorem 2: *For the two-level overlapping Schwarz method, using exact interior solvers, when the overlap is uniformly of width $O(\delta)$ then C_0^2 in Assumption 1 is bounded independently of h and grows linearly as $1 + H/\delta$.*

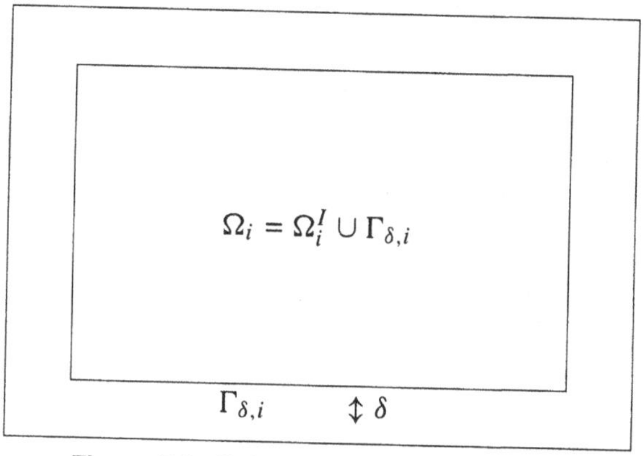

Figure 5.2. Strip along Subdomain Edge

SKETCH OF PROOF: Replace the bound $|\nabla\theta_i|^2_{L^\infty} \leq C/H^2$ by $|\nabla\theta_i|^2_{L^\infty} \leq C/\delta^2$. Then

$$||\theta_i - \bar{\theta}_i||^2_{L^\infty(K)} \leq C(h/\delta)^2. \tag{5.14}$$

As above

$$|\boldsymbol{u}_i|^2_{H^1(K)} \leq 2|\bar{\theta}_i \boldsymbol{w}|^2_{H^1(K)} + Ch^{-2}||I^h((\theta_i - \bar{\theta}_i)\boldsymbol{w})||^2_{L^2(K)}.$$

The last term above is identically zero for all elements K in the interior of an Ω_i^I. Therefore, when we take the sum over all the elements, the last term only includes those elements in the overlap regions,

$$\sum_{i=1}^{p} |\boldsymbol{u}_i|^2_{H^1(\Omega)} \leq C|\boldsymbol{w}|^2_{H^1(\Omega)} + \sum_{i=1}^{p} C\delta^{-2}||\boldsymbol{w}||^2_{L^2(\Gamma_{\delta,i})}.$$

To complete the proof, we now apply Lemma 6 together with (5.11) and (5.12).

□

The results of Theorems 1 and 2 may be extended to any second order, self-adjoint, uniformly elliptic PDE with general boundary conditions. They may also be extended to elliptic systems, such as the equations of linear elasticity. In addition, they may be extended to the case of nonnested coarse grids. They also apply when the local and coarse grid solves are replaced with spectrally equivalent forms.

For the extension to non-self-adjoint elliptic PDEs (which is much more difficult) see Section 5.4.2.

Notes and References for Section 5.3.1

- The analysis of additive overlapping Schwarz methods for the case of many subdomains is due to Dryja and Widlund [Dry89].

- Later, using a similar abstract framework to that presented in Section 5.2, Bramble, Pasciak, Wang, and Xu [Bra91b] restructured the Dryja-Widlund proof for the multiplicative Schwarz method.
- A proof for higher order p-version finite elements may be found in Pavarino [Pav94].
- The use of a partition of unity, in order to obtain upper bounds on C_0^2, is now common in many proofs for domain decomposition and multilevel algorithms.
- Lemma 6 may be found in Dryja and Widlund [Dry94b].
- Analysis for the case of a nonnested coarse grid may be found in X. Cai [Cai95] and Chan, Smith, and Zou [Cha94d].
- A statement and proof of a more general *inverse inequality* can be found in Theorem 3.2.6 in Ciarlet [Cia78].
- The bounds in the case when the coefficients of the elliptic operator are highly varying are still not completely understood.

5.3.2 Iterative Substructuring

In this section we provide an elementary proof for the convergence of the standard iterative substructuring algorithm given in Section 4.3.4. Since the algorithms and resulting convergence analysis are much more complex in three dimensions, we consider only a model problem in two dimensions. We do, however, allow large jumps in the coefficients between any two neighboring subdomains.

Theorem 3: *Consider the following problem in weak form: find $\boldsymbol{u} \in H_0^1(\Omega)$ such that*

$$a(\boldsymbol{u}, v) = \sum_i \int_{\Omega_i} \rho_i (\nabla \boldsymbol{u}) \cdot (\nabla v)\, dx = \int_{\Omega} f v\, dx \qquad v \in H_0^1(\Omega).$$

The $\rho_i > 0$ are constant in each Ω_i but may have arbitrary jumps between subdomains. For the standard iterative substructuring (see Algorithm 4.3.2) the condition number of the preconditioned system (using the abstract additive Schwarz preconditioner) is bounded independently of ρ_i and grows as $C(1 + \log(H/h))^2$.

The subspaces used in the theory (and the algorithm) are the coarse grid subspace, V^H; the local subspaces associated with the nonoverlapping regions, $V_i^h = V^h \cap H_0^1(\Omega_i)$; and the edge subspace, V_{jk}^h, defined in Section 4.6.

In the proof of this result, we construct the decomposition completely locally, one substructure at a time. The results in Section 4.3.4 then imply that the global bounds are no larger than the bounds obtained locally, since the preconditioner has the null space property. The construction is completely independent of the coefficients ρ_i. For simplicity we do the construction for an interior substructure. Boundary substructures are dealt with in a similar manner with special care to handle the boundary condition.

Let $\boldsymbol{u} \in V^h$ be any finite element function and consider its restriction to a single substructure, Ω_i. For the coarse grid component of the decomposition we must use the interpolant of $\boldsymbol{u}$, that is,

$$\boldsymbol{u}_0 = I^H \boldsymbol{u}.$$

This is because all functions associated with the edge and interior spaces are identically zero on the coarse grid nodes. Thus we must use the pointwise interpolation in the decomposition; we are not free to use the L^2-projection used in the previous proof.

Let $\mathcal{H}^j$ denote the discrete harmonic extension from edge Γ_j into the interior of Ω_i, see Section 4.6. Then define the components in the edge spaces by

$$\boldsymbol{u}_j = \mathcal{H}^j (\boldsymbol{u} - \boldsymbol{u}_0)|_{\Gamma_j} \qquad \text{for } j = 1, 2, 3.$$

The component in the interior space is given by

$$\boldsymbol{u}_4 = \boldsymbol{u} - \sum_{j=0}^{3} \boldsymbol{u}_j.$$

It is immediately obvious that $\boldsymbol{u}_4 \in V^h \cap H_0^1(\Omega_i)$ and $\boldsymbol{u} = \sum_{j=0}^{4} \boldsymbol{u}_j$. All that remains is to verify that the energy of each of these five pieces on Ω_i is bounded by the energy of $\boldsymbol{u}$.

For the boundedness of $\boldsymbol{u}_0$, we need the following discrete Sobolev type inequality. This says that in two dimensions, the maximum value of a finite element function grows relatively slowly with its energy.

Lemma 7: *For piecewise linear finite elements in two dimensions, let $\boldsymbol{u}_k$ be the value of $\boldsymbol{u}$ at any point in Ω_i; then*

$$||\boldsymbol{u} - \boldsymbol{u}_k||^2_{L^\infty(\Omega_i)} \le C(1 + \log(H/h))|\boldsymbol{u}|^2_{H^1(\Omega_i)}$$

and

$$||\boldsymbol{u}||^2_{L^\infty(\Omega_i)} \le C(1 + \log(H/h))||\boldsymbol{u}||^2_{H^1(\Omega_i)}.$$

For a proof see, for instance, Bramble [Bra66]. The second part of the lemma is not needed for the proof of Theorem 3, but is used in the proof of Theorem 5.

In addition, the following estimate, in two dimensions, is needed:

$$|\boldsymbol{u}_0|^2_{H^1(\Omega_i)} \le C(\boldsymbol{u}_{\max} - \boldsymbol{u}_{\min})^2.$$

This follows immediately since $\boldsymbol{u}_0$ is linear on Ω_i and its gradient is bounded by $(\boldsymbol{u}_{\max} - \boldsymbol{u}_{\min})/H$. Here $\boldsymbol{u}_{\max}$ and $\boldsymbol{u}_{\min}$ denote the extreme values of $\boldsymbol{u}$ in Ω_i.

For the edge spaces we construct a particular extension of $\boldsymbol{u}|_{\Gamma_j}$ that has bounded energy and conclude that since the discrete harmonic extension is the minimal energy extension (see Section 4.6), it must have the same (or a better) bound. Let ϑ_j be a continuous, piecewise linear function that is identically 1 on the finite element nodes

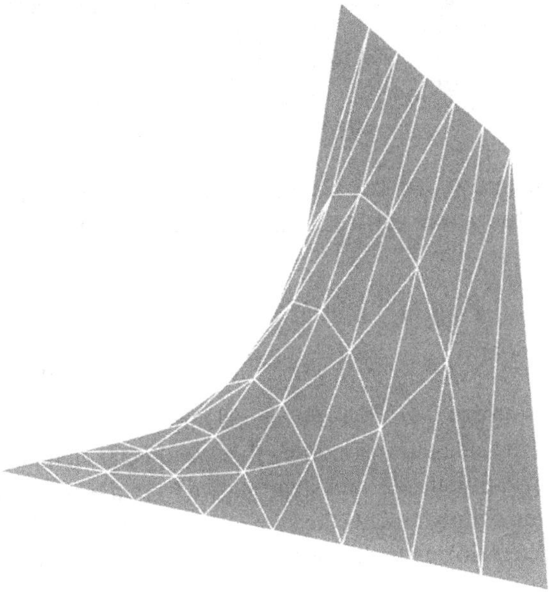

Figure 5.3. Partition Function ϑ_j

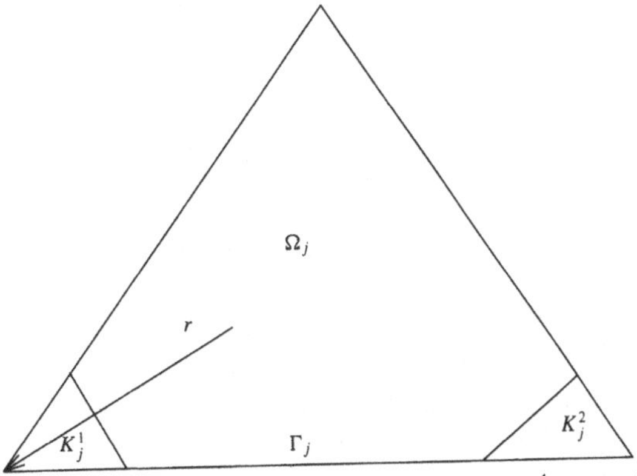

Figure 5.4. Location of Special Elements K_j^1 and K_j^2

of Γ_j and 0 on the finite element nodes of the other two edges. The value ϑ_j at the substructure vertices is also zero, see Figure 5.3. Let

$$\tilde{\boldsymbol{u}}_j = I^h(\vartheta_j(\boldsymbol{u} - \boldsymbol{u}_0)).$$

As we learned from Theorem 1 the issue is what bounds can be given on the gradient of ϑ_j. For the elements, denoted by K_j^1 and K_j^2, adjacent to the endpoints of Γ_j the gradient may be bounded by C/h. For all other elements of Ω_i it can be bounded by C/r where r is the distance to the nearest endpoint of Γ_j; see Figure 5.4.

The next step is to bound the energies of the parts of $\tilde{\boldsymbol{u}}_j$ associated with K_j^l.

$$
\begin{aligned}
|\tilde{\boldsymbol{u}}_j|^2_{H^1(K_j^l)} &\leq \frac{(\boldsymbol{u}_{\max} - \boldsymbol{u}_{\min})^2}{h^2} h^2, \\
&\leq C(1 + \log(H/h))|\boldsymbol{u}|^2_{H^1(\Omega_i)}.
\end{aligned}
$$

For the other elements we use the fact that the gradient, bounded by C/r, $r > h$, is only large near the substructure vertices. On each element first separate out the average of ϑ_j,

$$
\begin{aligned}
|\tilde{\boldsymbol{u}}_j|^2_{H^1(K)} &= |I^h(\vartheta_j(\boldsymbol{u} - \boldsymbol{u}_0))|^2_{H^1(K)} \\
&\leq 2|\bar{\vartheta}_j(\boldsymbol{u} - \boldsymbol{u}_0)|^2_{H^1(K)} + 2|I^h(\bar{\vartheta}_j - \vartheta_j)(\boldsymbol{u} - \boldsymbol{u}_0)|^2_{H^1(K)}.
\end{aligned}
$$

The first term creates no problems but the second term must be dealt with carefully.

Now sum over all the elements, excluding the two special elements. In the first step apply the inverse inequality; the second step follows from the bound on the gradient of ϑ_j, while the third step simply bounds the sum by an integral in polar coordinates. The origin for the polar coordinates is the closest coarse grid node. The last step follows from Lemma 7.

$$
\begin{aligned}
2 \sum_{K \subset \Omega_i \setminus K_j^l} |I^h(\bar{\vartheta}_j - \vartheta_j)(\boldsymbol{u} - \boldsymbol{u}_0)|^2_{H^1(K)} &\leq 2 \sum_{K \subset \Omega_i \setminus K_j^l} Ch^{-2} ||I^h(\bar{\vartheta}_j - \vartheta_j)(\boldsymbol{u} - \boldsymbol{u}_0)||^2_{L^2(K)}, \\
&\leq Cr^{-2} \sum_{K \subset \Omega_i \setminus K_j^l} ||(\boldsymbol{u} - \boldsymbol{u}_0)||^2_{L^2(K)}, \\
&\leq C \int_{r=h}^{H} \int_{\theta} r^{-2} r (\boldsymbol{u}_{\max} - \boldsymbol{u}_{\min})^2 \, d\theta \, dr, \\
&\leq C(1 + \log(H/h))^2 |\boldsymbol{u}|^2_{H^1(\Omega_i)}.
\end{aligned}
$$

A more detailed construction of this type is carried out for tetrahedra in three dimensions in Dryja, Smith, and Widlund [Dry94a].

The main features distinguishing this proof from the one of Theorem 1 are (a) the fact that we must use the interpolant of $\boldsymbol{u}$ for the coarse grid component and (b) the gradient of ϑ_j is not uniformly bounded, but has the singular, C/r, behavior. Each of (a) and (b) accounts for one of the two powers of $(1 + \log(H/h))$ in the bound. The difference in behavior in two and three dimensions is also caused by (a), since the version of Lemma 7 for three dimensions, given by Lemma 8 below, is much weaker.

Lemma 8: *For piecewise linear finite elements in three dimensions, let $\boldsymbol{u}_k$ be the value of $\boldsymbol{u}$ at any point in Ω_i; then*

$$
||\boldsymbol{u} - \boldsymbol{u}_k||^2_{L^\infty(\Omega_i)} \leq C\frac{H}{h}|\boldsymbol{u}|^2_{H^1(\Omega_i)},
$$

and

$$
||\boldsymbol{u}||^2_{L^\infty(\Omega_i)} \leq C\frac{H}{h}||\boldsymbol{u}||^2_{H^1(\Omega_i)}.
$$

In three dimensions one of the $(1+\log(H/h))$ terms must be replaced, in the statement of Theorem 3, with an (H/h) term.

PROOF OF THEOREM 3: To calculate the value of C_0^2 in Assumption 1 construct the following decomposition of $\boldsymbol{u}$:

$$\begin{aligned}
\boldsymbol{u}_0 &= I^H \boldsymbol{u}, \\
\boldsymbol{u}_j &= \mathcal{H}^j(\boldsymbol{u} - \boldsymbol{u}_0)|_{\Gamma_j} \qquad \text{for } j = 1, 2, 3, \\
\boldsymbol{u}_4 &= \boldsymbol{u} - \sum_{j=0}^{3} \boldsymbol{u}_j.
\end{aligned}$$

The energy of the first term may be bounded by Lemma 7

$$\begin{aligned}
|\boldsymbol{u}_0|^2_{H^1(\Omega_i)} &\leq C(\boldsymbol{u}_{\max} - \boldsymbol{u}_{\min})^2, \\
&\leq C(1 + \log(H/h))|\boldsymbol{u}|^2_{H^1(\Omega_i)}.
\end{aligned}$$

For the edge spaces

$$\begin{aligned}
|\boldsymbol{u}_j|^2_{H^1(\Omega_i)} &\leq |\vartheta_j(\boldsymbol{u} - \boldsymbol{u}_0)|^2_{H^1(\Omega_i)}, \\
&\leq C\sum_l |\vartheta_j(\boldsymbol{u} - \boldsymbol{u}_0)|^2_{H^1(K_j^l)} + C\sum_{K\subset\Omega_i\setminus K_j^l} |\bar{\vartheta}_j(\boldsymbol{u} - \boldsymbol{u}_0)|^2_{H^1(K)} \\
&\quad + C\sum_{K\subset\Omega_i\setminus K_j^l} |I^h(\bar{\vartheta}_j - \vartheta_j)(\boldsymbol{u} - \boldsymbol{u}_0)|^2_{H^1(K)}.
\end{aligned}$$

Next apply an inverse inequality to the last term.

$$\begin{aligned}
\sum_{K\subset\Omega_i\setminus K_j^l} |I^h(\bar{\vartheta}_j - \vartheta_j)(\boldsymbol{u} - \boldsymbol{u}_0)|^2_{H^1(K)} &\leq \sum_{K\subset\Omega_i\setminus K_j^l} h^{-2}||I^h(\bar{\vartheta}_j - \vartheta_j)(\boldsymbol{u} - \boldsymbol{u}_0)||^2_{L^2(K)}, \\
&\leq Cr^{-2}\sum_{K\subset\Omega_i\setminus K_j^l} ||(\boldsymbol{u} - \boldsymbol{u}_0)||^2_{L^2(K)}, \\
&\leq C\int_{r=h}^{H}\int_{\theta} r^{-2} r(\boldsymbol{u}_{\max} - \boldsymbol{u}_{\min})^2\, d\theta\, dr, \\
&\leq C(1 + \log(H/h))^2|\boldsymbol{u}|^2_{H^1(\Omega_i)}.
\end{aligned}$$

By combining the decompositions for each subdomain we obtain a decomposition of $\boldsymbol{u}$ in Ω that satisfies Assumption 1 independently of the coefficients ρ_i.

The value of ω is trivially 1. The value of $\rho(\mathcal{E}) \leq 4$ since at most four local subspaces overlap. □

Notes and References for Section 5.3.2

- The proof given here is based on the proof in Dryja and Widlund [Dry89]. See also Dryja, Smith and Widlund [Dry94a] for a similar, though more complicated analysis in three dimensions.
- A more complicated proof for a slightly different but related algorithm is given in Bramble, Pasciak, and Schatz [Bra86].
- For analysis of iterative substructuring in three dimensions for spectral methods, see Pavarino and Widlund [Pav94a], [Pav94b].

5.3.3 Neumann-Neumann

In this section we provide a proof of the convergence for the Neumann-Neumann algorithm with and without a coarse grid problem. For these algorithms, the subspaces (see Section 4.6) are the interior subspaces

$$V^h \cap H_0^1(\Omega_i),$$

and the subdomain boundary subspaces given by

$$\tilde{V}_i^h = \{\boldsymbol{u} \in \tilde{V}^h | \boldsymbol{u}^h(x_i) = 0, \ \forall x_i \in \cup_j \partial\Omega_j \setminus \partial\Omega_i\}.$$

The x_i denote only the finite element nodes on the interface $\cup_j \partial\Omega_j$. The space $\tilde{V}^h$ is the subspace of V^h of those functions that are discrete harmonic in the interiors of Ω_i. Note that they are discrete harmonic as calculated by using the usual inner product, $a(\cdot, \cdot)$. The bilinear form that we will consider is given by

$$a(\boldsymbol{u}, \boldsymbol{v}) = \sum_i \int_{\Omega_i} (\nabla \boldsymbol{u}) \cdot (\nabla \boldsymbol{v}) \, dx.$$

More general self-adjoint, second order elliptic problems may also be considered without difficulty.

The boundary subspaces require a careful choice of the local solvers; see Section 4.6. This is for two reasons: (1) to deal with the null space of interior Neumann boundary value problems and (2) to scale the contribution at each node by the number of subdomains that a node is shared by. On the boundary spaces $\tilde{V}_i^h$, we use the complex-looking approximate solver,

$$a_i(\boldsymbol{u}, \boldsymbol{v}) = \hat{a}_{\Omega_i}(\hat{\mathcal{H}}_i I^h(\lambda_i \boldsymbol{u}), \hat{\mathcal{H}}_i I^h(\lambda_i \boldsymbol{v})).$$

The λ_i scales the value of $\boldsymbol{u}$ at each finite element node by the number of subdomains the node is contained in; see Section 4.3.2. The $\hat{\mathcal{H}}_i$ is the discrete harmonic extension into Ω_i using the $\hat{a}_{\Omega_i}(\cdot, \cdot)$ inner product, defined below. Note that, depending on $\hat{a}_{\Omega_i}(\cdot, \cdot)$, this extension may not be the same as the usual discrete harmonic extension. On a subdomain that touches a boundary on which Dirichlet conditions are applied, one may use

$$\hat{a}_{\Omega_i}(\boldsymbol{u}, \boldsymbol{v}) = a_{\Omega_i}(\boldsymbol{u}, \boldsymbol{v}).$$

However, for an interior subdomain the resulting operator would be singular, since $a_{\Omega_i}(1, 1) = 0$. The way this is commonly dealt with is to introduce a zero order term that removes the singularity:

$$\hat{a}_{\Omega_i}(\boldsymbol{u}, \boldsymbol{v}) = a_{\Omega_i}(\boldsymbol{u}, \boldsymbol{v}) + H^{-2} \int_{\Omega_i} \boldsymbol{uv}\, dx.$$

The matrix representation of the local correction is, perhaps, clearer,

$$P_i = D\tilde{R}_i^T \hat{S}_i^{-1} \tilde{R}_i DS.$$

Recall from Section 4.3.2 that D is the diagonal scaling matrix and here $\tilde{R}_i$ is the restriction operator that extracts all the coefficients associated with boundary of subdomain Ω_i. The matrix $\hat{S}_i$ is the Schur complement associated with the inner product $\hat{a}_{\Omega_i}(\cdot, \cdot)$ given above.

For the balancing Neumann-Neumann method we need the additional "piecewise constant" coarse grid subspace defined by

$$V_0^h = \text{span}_i\{\mathcal{H}I^h \lambda_i^{-1}\}.$$

This space, in fact, does not contain the piecewise constants, but approximations to them that are continuous and thus in V^h.

Theorem 4: *For the one level Neumann-Neumann algorithm (see Section 4.3.1) the condition number of the preconditioned problem may be bounded by*

$$\frac{C(1 + \log(H/h))^2}{H^2}.$$

In particular C_0^2 in Assumption 1 is bounded by CH^{-2}, $\rho(\mathcal{E}) \le C$, and ω in Assumption 3 is bounded by $C(1 + \log(H/h))^2$.

Note that for this algorithm the ω is in general greater than 2. This is not a problem since the Neumann-Neumann method is always used as an additive method.

Theorem 5: *For the balancing Neumann-Neumann algorithm (see Section 4.3.3) the condition number of the preconditioned problem may be bounded by*

$$C(1 + \log(H/h))^2,$$

when the scaling $(1 + \log(H/h))^2$ is applied to the coarse grid problem.

DISCUSSION AND PROOF OF THEOREM 4: Define the partition

$$\tilde{\boldsymbol{u}}_i = \mathcal{H}I^h(\lambda_i^{-1}\boldsymbol{u})$$

and

$$\boldsymbol{u}_i = \left(\boldsymbol{u} - \sum_j \tilde{\boldsymbol{u}}_j\right)\Big|_{\Omega_i}.$$

Recall that λ_i equals the number of subdomains whose closure contains the interface node x_i. The $\boldsymbol{u}_i$ are extended by zero outside Ω_i, while the $\tilde{\boldsymbol{u}}_i$ are extended onto the interface by zero and into the interiors of the neighboring subdomains as discrete harmonic functions. By construction $\boldsymbol{u} = \sum_i(\boldsymbol{u}_i + \tilde{\boldsymbol{u}}_i)$ and the $\boldsymbol{u}_i$ are zero on the substructure boundaries. Now, using the fact that since $\tilde{\boldsymbol{u}}_i$ is discrete harmonic, its energy on Ω_i is bounded by the energy of $\boldsymbol{u}$ and the definition of $a_i(\cdot, \cdot)$,

$$\begin{aligned}
\sum_i a_i(\tilde{\boldsymbol{u}}_i, \tilde{\boldsymbol{u}}_i) &\leq C \sum_i \hat{a}_i(\boldsymbol{u}, \boldsymbol{u}), \\
&= C[\sum_i a_{\Omega_i}(\boldsymbol{u}, \boldsymbol{u}) + H^{-2} \int_{\Omega_i} \boldsymbol{u}^2\, dx], \\
&\leq C[a(\boldsymbol{u}, \boldsymbol{u}) + H^{-2}||\boldsymbol{u}||^2_{L^2(\Omega)}], \\
&< CH^{-2}[a(\boldsymbol{u}, \boldsymbol{u}) + ||\boldsymbol{u}||^2_{L^2(\Omega)}], \\
&< CH^{-2}a(\boldsymbol{u}, \boldsymbol{u}).
\end{aligned}$$

The final inequality follows from Friedrichs' inequality,

$$|\boldsymbol{u}|^2_{H^1(\Omega)} + ||\boldsymbol{u}||^2_{L^2(\Omega)} \leq Ca(\boldsymbol{u}, \boldsymbol{u}).$$

Note that the H^{-2} term is due to the constant functions. When they are provided with their own subspace in the balancing Neumann-Neumann method they are no longer troublesome.

To estimate the other terms, we use the mutual orthogonality of the various $\boldsymbol{u}_i$,

$$\begin{aligned}
\sum_i a(\boldsymbol{u}_i, \boldsymbol{u}_i) &= a(\sum_i \boldsymbol{u}_i, \sum_j \boldsymbol{u}_j), \\
&= a(\boldsymbol{u} - \sum_i \tilde{\boldsymbol{u}}_i, \boldsymbol{u} - \sum_j \tilde{\boldsymbol{u}}_j), \\
&\leq 2a(\boldsymbol{u}, \boldsymbol{u}) + 2a(\sum_i \tilde{\boldsymbol{u}}_i, \sum_j \tilde{\boldsymbol{u}}_j), \\
&\leq 2a(\boldsymbol{u}, \boldsymbol{u}) + C \sum_i a(\tilde{\boldsymbol{u}}_i, \tilde{\boldsymbol{u}}_i), \\
&\leq CH^{-2}a(\boldsymbol{u}, \boldsymbol{u}).
\end{aligned}$$

Applying the same type of coloring argument as in the proof of Theorem 1, one easily obtains the bound $\rho(\mathcal{E}) \leq N_c$.

It remains to estimate the ω in Assumption 3, $a(\boldsymbol{u}, \boldsymbol{u}) \leq \omega a_i(\boldsymbol{u}, \boldsymbol{u})$, $\forall \boldsymbol{u} \in \tilde{V}_i^h$. Let $\boldsymbol{u} \in \tilde{V}_i^h$. We need to bound $a(\boldsymbol{u}, \boldsymbol{u})$ in terms of $a_i(\boldsymbol{u}, \boldsymbol{u})$. Since $\boldsymbol{u}$ is identically zero in all subdomains that do not touch Ω_i we have (see Figure 5.5)

$$a(\boldsymbol{u}, \boldsymbol{u}) = a_{\Omega_i}(\boldsymbol{u}, \boldsymbol{u}) + \sum_{\bar{\Omega}_j \cap \bar{\Omega}_i \neq \emptyset} a_{\Omega_j}(\boldsymbol{u}, \boldsymbol{u}). \tag{5.15}$$

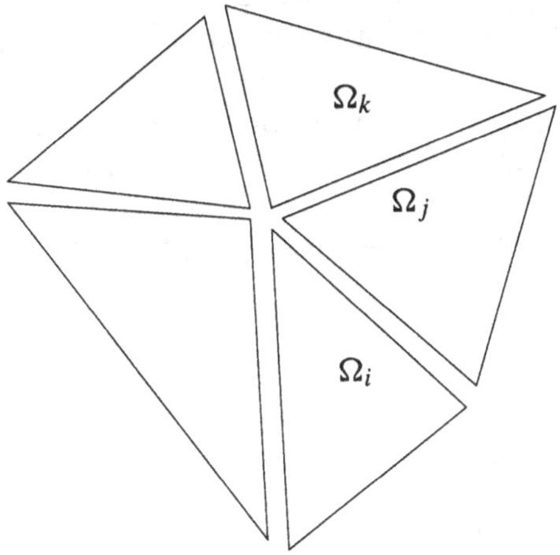

Figure 5.5. Substructure Ω_j with Neighbors

Thus each term on the right hand side must be bounded by $a_i(\boldsymbol{u}, \boldsymbol{u})$. To simplify the technical content we consider only problems in two dimensions.

First consider the term $a_{\Omega_i}(\boldsymbol{u}, \boldsymbol{u})$. Decompose $\boldsymbol{u}$ into two parts, one of which is zero on the edges,

$$\boldsymbol{v} = \tfrac{1}{2}\mathcal{H}_i I^h((2 - \lambda_i)\boldsymbol{u})$$

and the remainder

$$\boldsymbol{w} = \boldsymbol{u} - \boldsymbol{v} = \tfrac{1}{2}\mathcal{H}_i I^h(\lambda_i \boldsymbol{u}).$$

Now

$$a_{\Omega_i}(\boldsymbol{w}, \boldsymbol{w}) \leq \tfrac{1}{4}a_i(\boldsymbol{u}, \boldsymbol{u}).$$

Thus

$$\begin{aligned} a_{\Omega_i}(\boldsymbol{u}, \boldsymbol{u}) &\leq 2a_{\Omega_i}(\boldsymbol{w}, \boldsymbol{w}) + 2a_{\Omega_i}(\boldsymbol{v}, \boldsymbol{v}), \\ &\leq \tfrac{1}{2}a_i(\boldsymbol{u}, \boldsymbol{u}) + 2a_{\Omega_i}(\boldsymbol{v}, \boldsymbol{v}). \end{aligned}$$

It remains to bound the second term. Note that $\boldsymbol{v}$ is nonzero at the substructure vertices, is zero on the edges and has an energy minimizing extension into Ω_i. Therefore it has smaller energy than the zero extension from the substructure vertices.

$$\begin{aligned} a_{\Omega_i}(\boldsymbol{v}, \boldsymbol{v}) &\leq C(a_{\Omega_i}(I^h(\phi_1^i \boldsymbol{u}), I^h(\phi_1^i \boldsymbol{u})) + a_{\Omega_i}(I^h(\phi_2^i \boldsymbol{u}), I^h(\phi_2^i \boldsymbol{u})) \\ &\quad + a_{\Omega_i}(I^h(\phi_3^i \boldsymbol{u}), I^h(\phi_3^i \boldsymbol{u}))), \\ &\leq C\|\hat{\mathcal{H}}_i I^h(\lambda_i^{-1}\boldsymbol{u})\|_{L^\infty(\Omega_i)}^2, \\ &\leq C(1 + \log(H/h))\|\hat{\mathcal{H}}_i I^h(\lambda_i^{-1}\boldsymbol{u})\|_{H^1(\Omega_i)}^2, \\ &\leq C(1 + \log(H/h))a_i(\boldsymbol{u}, \boldsymbol{u}). \end{aligned}$$

Here we have used Lemma 7; this introduces a $(1 + \log(H/h))$ factor.

Let Ω_j denote a substructure that shares a common edge with Ω_i; see Figure 5.5. From (5.15), we need to bound $a_{\Omega_j}(\boldsymbol{u}, \boldsymbol{u})$ by $a_i(\boldsymbol{u}, \boldsymbol{u})$.

Let θ_{ij} be a function that is one on the finite element nodes on the interior of the common edge shared by Ω_i and Ω_j, and zero on the rest of the finite element nodes on the skeleton. Note that θ_{ij} is exactly zero at the substructure vertices (the coarse grid nodes). Decompose $\boldsymbol{u}$ into two parts

$$\boldsymbol{v} = \mathcal{H}_j I^h((1 - \theta_{ij})\boldsymbol{u})$$

and

$$\boldsymbol{w} = \tfrac{1}{2}\mathcal{H}_j I^h(\theta_{ij}\lambda_i \boldsymbol{u}) = \mathcal{H}_j I^h(\theta_{ij}\boldsymbol{u}).$$

Note the similarity to the decomposition used above. Again, $\boldsymbol{v}$ is zero on the common edge, but, in addition, $\boldsymbol{w}$ is zero at the endpoints of the common edge.

$$a_{\Omega_j}(\boldsymbol{u}, \boldsymbol{u}) \le 2a_{\Omega_j}(\boldsymbol{v}, \boldsymbol{v}) + 2a_{\Omega_j}(\boldsymbol{w}, \boldsymbol{w}).$$

Now the first term involving $\boldsymbol{v}$ may be dealt with in exactly the same way as previously. The second term is more painful. Recall that in the proof of Theorem 3 we showed that the energy of an element in an edge subspace $\boldsymbol{u}_j = \mathcal{H}_j(\boldsymbol{u} - \boldsymbol{u}_0)$, $j = 1, 2, 3$ could be bounded by $C(1 + \log(H/h))^2 |\boldsymbol{u}|^2_{H^1(\Omega_i)}$. Using the same techniques it can be shown that

$$|\boldsymbol{w}|^2_{H^1(\Omega_i)} \le C(1 + \log(H/h))^2 ||\hat{\mathcal{H}} I^h(\lambda_i \boldsymbol{u})||^2_{H^1(\Omega_i)}.$$

To complete the required bound we observe that $||\hat{\mathcal{H}} I^h(\lambda_i \boldsymbol{u})||^2_{H^1(\Omega_i)}$ may be bounded by $a_i(\boldsymbol{u}, \boldsymbol{u})$ because of the $H^{-2}||\boldsymbol{u}||^2_{L^2(\Omega_i)}$ term in the definition of $a_i(\cdot, \cdot)$.

Similar calculations may be carried out for a domain that shares a single vertex with Ω_i, for instance, Ω_k in Figure 5.5.

□

DISCUSSION AND PROOF OF THEOREM 5: The difference between this proof and the one above is the inclusion of the coarse grid component in the decomposition of $\boldsymbol{u}$. The component in the coarse grid space in the decomposition is obtained by first constructing a basis for the space V_0. Define the basis functions

$$\mathcal{H} I^h \lambda_i^{-1}.$$

These functions are discrete harmonic in the interiors of the subdomains and nonzero only on $\partial\Omega_i$ and the portion of $\cup_i \partial\Omega_j$ less than one element away from $\partial\Omega_i$. Let $\bar{\boldsymbol{u}}_i$ be the average of $\boldsymbol{u}$ on Ω_i. Then the coarse grid component is given by

$$\boldsymbol{u}_0 = \sum_i \bar{\boldsymbol{u}}_i \mathcal{H} I^h \lambda_i^{-1}.$$

Here, to simplify the notation, we assume that all of the substructures that intersect the boundary have at least one face on the boundary. The more general, trickier case is handled in Dryja and Widlund [Dry95]. The contributions from the local subspaces are defined similarly as in the proof above by

$$\tilde{u}_i = \mathcal{H}I^h(\lambda_i^{-1}(u - u_0))$$

and

$$u_i = (u - \sum_j \tilde{u}_j - u_0)|_{\Omega_i}.$$

As in the proof of Theorem 4,

$$\begin{aligned}\sum_i a_i(\tilde{u}_i, \tilde{u}_i) &\le \sum_i C\hat{a}_i(u - \bar{u}_i, u - \bar{u}_i),\\ &\le C\Big(a(u,u) + \sum_i \frac{1}{H^2}||u - \bar{u}_i||^2_{L^2(\Omega_i)}\Big),\\ &\le Ca(u,u),\end{aligned}$$

by a Poincaré inequality.

We next estimate

$$\begin{aligned}a_0(u_0, u_0) &= (1 + \log(H/h))^{-2} a(u_0, u_0),\\ &= (1 + \log(H/h))^{-2} a(u - \sum_j u_j - \sum_i \tilde{u}_i, u - \sum_j u_j - \sum_i \tilde{u}_i),\\ &\le C(1 + \log(H/h))^{-2}(a(u,u) + \sum_i a(\tilde{u}_i, \tilde{u}_i)),\\ &\le C(1 + \log(H/h))^{-2}(a(u,u) + C(1 + \log(H/h))^2 \sum_i a_i(\tilde{u}_i, \tilde{u}_i)),\\ &\le Ca(u,u).\end{aligned}$$

From the definition of $a_0(\cdot, \cdot)$, it is obvious that

$$a(u,u) \le C(1 + \log(H/h))^2 a_0(u,u) \qquad \forall u \in V_0. \qquad \square$$

Notes and References for Section 5.3.3

- The first proof we are aware of for the Neumann-Neumann algorithm, using a different scaling, was given by Dryja and Widlund [Dry90], [Dry92]. Full details of the proof in three dimensions are given in Dryja and Widlund [Dry95].
- Later De Roeck and Le Tallec [De R91] provided some analysis with the bound given in Theorem 4.
- Mandel [Man93] first proved the result for the balancing algorithm. See also, Dryja and Widlund [Dry95] and Dryja, Smith, and Widlund [Dry94a] for analysis of several variants of the algorithm.

5.3.4 Multilevel Diagonal Scaling

In this section we provide a proof of the optimality of the multilevel diagonal scaling method. This type of proof may be used for both the additive and the multiplicative form (which is simply V-cycle multigrid with a Jacobi smoother; see Section 3.8). The proof can also be easily modified for the BPX method (see Section 3.1.2).

As in Chapter 3 assume that the domain Ω has been triangulated with a nested family of triangles (or tetrahedrals). The characteristic diameter of elements on level i is denoted by $h^{(i)}$. The coarsest level has elements of size $h^{(0)}$ and the finest has elements of size $h^{(J-1)}$. (Note that we used $J+1$ levels in Chapter 3; this change is only for convenience of notation and should cause no confusion.)

The subspaces for the multilevel algorithm considered here are given by

$$V_l^i = \text{span}\{\phi_l^i\},$$

where ϕ_l^i is the lth finite element basis function on the ith level. For the more general multilevel overlapping Schwarz methods $V_l^i = H_0^1(\Omega_l^i) \cap V^h$.

Theorem 6: *For the multilevel diagonal scaling method, Algorithm 3.1.2 (more generally for multilevel overlapping Schwarz methods, e.g., Algorithm 3.1.1), the condition number of the preconditioned problem may be bounded independently of $h^{(0)}$ and the number of levels J.*

In fact, this theorem may be generalized to include some standard V-cycle multigrid methods by using the result demonstrated in Section 3.2.1 that certain recursive multilevel methods are identical in exact arithmetic to their nonrecursive multiplicative counterparts. Of course, the multiplicative algorithms may be analyzed in the same framework as their additive counterparts.

To give an elementary proof we restrict attention to two dimensions and use a regular uniform triangulation; that is, each element on level i is decomposed into four elements on level $i+1$. With such a decomposition

$$\frac{h_k^{(i)}}{h_l^{(i-1)}} = \gamma = \frac{1}{2},$$

where element k is any of the four subelements of element l.

For the multilevel diagonal scaling method, associated with each finite element node, l, on each level, i, is the one-dimensional subspace

$$V_l^i = \text{span}\{\phi_l^i\}.$$

Let O_l^i denote the support of ϕ_l^i, and, in addition, let V^i denote the union of all the subspaces associated with level i, that is, $V^i = \cup_l V_l^i$.

DISCUSSION AND PROOF OF THEOREM 6: We must derive a bound for C_0^2 in Assumption 1. To simplify the proof we only consider a convex domain. Nonconvex domains may be dealt with by imbedding them into a larger convex domain, but the proof becomes exceedingly complicated. Therefore, we assume Ω is convex.

We need to construct a decomposition of $\boldsymbol{u}$ into the subspaces V_l^i. First decompose $\boldsymbol{u}$ into the various levels by using the $a(\cdot,\cdot)$ projections, denoted by $\boldsymbol{P}_i$.

$$\begin{aligned} \boldsymbol{u}^0 &= \boldsymbol{P}_0\boldsymbol{u}, \\ \boldsymbol{u}^1 &= (\boldsymbol{P}_1 - \boldsymbol{P}_0)\boldsymbol{u}, \\ &\dots \\ \boldsymbol{u}^{J-2} &= (\boldsymbol{P}_{J-2} - \boldsymbol{P}_{J-3})\boldsymbol{u}, \\ \boldsymbol{u}^{J-1} &= (\boldsymbol{I} - \boldsymbol{P}_{J-2})\boldsymbol{u}. \end{aligned}$$

The functions on each level must now be decomposed into the individual subspaces V_l^i. This is done by simply selecting the value of $\boldsymbol{u}^i$ at the nodes on that level,

$$\boldsymbol{u}_l^i = \boldsymbol{u}^i|_{x_l}\phi_l^i = I^{h^{(i)}}(\phi_l^i \boldsymbol{u}^i).$$

For the proof of the more general multilevel overlapping Schwarz method (which we do not give) the values of $\boldsymbol{u}_l^i$ for each subdomain are selected by using the same type of partitioning function as in the proof of Theorem 1.

It follows immediately from the definition of $\boldsymbol{u}_l^i$ that they form a valid decomposition of $\boldsymbol{u}$, that is,

$$\boldsymbol{u} = \sum_i \sum_l \boldsymbol{u}_l^i.$$

It is left to prove that this is a stable splitting. We have

$$\begin{aligned} \sum_l a(\boldsymbol{u}_l^i, \boldsymbol{u}_l^i) &\le \sum_l C|\boldsymbol{u}_l^i|^2_{H^1(\Omega)}, \\ &= \sum_l C|I^{h^{(i)}}(\phi_l^i \boldsymbol{u}^i)|^2_{H^1(O_l^i)}, \\ &\le \sum_l C(|\boldsymbol{u}^i|^2_{H^1(O_l^i)} + |\nabla\phi_l^i|^2_\infty ||\boldsymbol{u}^i||^2_{L^2(O_l^i)}), \\ &\le \sum_l C(|\boldsymbol{u}^i|^2_{H^1(O_l^i)} + (1/h^{(i)})^2 ||\boldsymbol{u}^i||^2_{L^2(O_l^i)}), \\ &\le C(|\boldsymbol{u}^i|^2_{H^1(\Omega)} + (1/h^{(i)})^2 ||\boldsymbol{u}^i||^2_{L^2(\Omega)}). \end{aligned}$$

This follows from the product rule for gradients and the fact that $|\nabla\phi_l^i| = O(1/h^{(i)})$.

It is now necessary to estimate $||\boldsymbol{u}^i||_{L^2(\Omega)}$ in terms of the H^1 seminorm. To do this we use the Aubin-Nitsche theory (Nitsche's trick); see, for example Ciarlet [Cia78], Section 3.2. This theory can be used to show that for a convex region, if $\boldsymbol{w} \in V^{h^{(i)}}$ then

$$||\boldsymbol{w} - \boldsymbol{P}_i\boldsymbol{w}||_{L^2(\Omega)} \le Ch^{(i)}|\boldsymbol{w}|_{H^1(\Omega)}.$$

Therefore, since $u^i = (I - P_{i-1})u^i$,

$$\|u^i\|_{L^2(\Omega)} \leq Ch^{(i-1)}|u^i|_{H^1(\Omega)}.$$

Combining this with the previous formula gives

$$\sum_l a(u_l^i, u_l^i) \leq C|u^i|^2_{H^1(\Omega)}.$$

The required result follows from the $a(\cdot,\cdot)$ orthogonality of the spaces V^i.

$$\begin{aligned}\sum_i \sum_l a(u_l^i, u_l^i) &\leq C \sum_i |u^i|^2_{H^1(\Omega)}, \\ &= C|u|^2_{H^1(\Omega)}, \\ &\leq Ca(u, u).\end{aligned}$$

Thus we have verified Assumption 1.

We next show how one may naively obtain the (less than sharp) bound $\rho(\mathcal{E}) \leq CJ$. Observe that on each level the nodes may be colored so that neighboring nodes are always of a different color. The number of colors needed for a regular mesh is a bounded constant; call it N_c. All the subspaces on a particular level of the same color may be combined to form a new subspace. Thus the total number of merged subspaces, and hence interfering projections, is $N_c J$. This gives us a crude upper bound on $\rho(\mathcal{E})$. Note that when analyzing multilevel methods there is no reason to exclude the coarsest subspace in Assumption 2 as was done for the two level methods.

To obtain a sharper bound we must be very careful and obtain explicit bounds on the individual components of $\mathcal{E}$. The proof is quite delicate and requires a careful set of steps between various matrix and vector norms. Straightforward calculations do not produce a sharp bound.

Let $u \in V_l^i$ and $v \in V_k^j$, $j \geq i$; then by the usual Cauchy-Schwarz inequality,

$$a(u, v) \leq |u|_{H^1(O_l^i \cap O_k^j)} |v|_{H^1(O_l^i \cap O_k^j)}.$$

But the diameter of $O_l^i \cap O_k^j$ is at most the size of O_k^j, which is $O(h^{(j)})$. Thus since u is piecewise linear on $O_l^i \cap O_k^j$,

$$|u|_{H^1(O_l^i \cap O_k^j)} \leq C\gamma^{j-i}|u|_{H^1(\Omega)}.$$

This follows by simply considering the fraction of the area of O_l^i that is covered by O_k^j. Using this results in

$$a(u, v) \leq C\gamma^{|j-i|}a(u, u)^{1/2}a(v, v)^{1/2} \qquad \forall u \in V_l^i,\ \forall v \in V_k^j.$$

That is, the farther apart the two levels are the more orthogonal are the resulting subspaces.

It is crucial to note that, in addition, if O_k^j is completely contained in a single coarser grid element that defines O_l^i (this is explained below) or is disjoint from O_l^i, then $a(u, v) = 0$. Because of this the matrix $\mathcal{E}$ is actually very sparse. In fact, if row I of $\mathcal{E}$ corresponds to the space V_l^i, then **(F1)** the number of nonzero columns associated with level j is bounded by $C\gamma^{-|i-j|}$. In addition, **(F2)** the number of nonzero rows, for any column, associated with any level is bounded by the number of colors.

Because it is so crucial to the derivation of the upper bound, we clarify why $a(\boldsymbol{u}, \boldsymbol{v}) = 0$ when O_k^j is completely contained in a single coarser grid element, K, that defines O_l^i. For piecewise linear finite elements $\nabla \boldsymbol{u}$ is constant over K, thus

$$\begin{aligned} a_K(\boldsymbol{u}, \boldsymbol{v}) &= \int_K \nabla \boldsymbol{u} \cdot \nabla \boldsymbol{v}\, dx, \\ &= \int_K C \cdot \nabla \boldsymbol{v}\, dx, \\ &= C \int_{\partial K} \boldsymbol{v}\, dx, \end{aligned}$$

by a Green's formula. But $\boldsymbol{v}$ is identically zero on the boundary of K since $O_k^j \subset K$. Hence $a_K(\boldsymbol{u}, \boldsymbol{v}) = 0$. For higher order finite elements the result $a(\boldsymbol{u}, \boldsymbol{v}) = 0$ is no longer valid. However, it can be shown that $a(\boldsymbol{u}, \boldsymbol{v})$ is "small" enough to prove that $\rho(\mathcal{E})$ is uniformly bounded. Thus the general result that the farther apart the two levels are, the more orthogonal are the associated subspaces, still holds, even when higher order finite element methods are used.

What remains to be shown is that the spectral radius of $\mathcal{E}$ can be bounded independently of the number of levels. This is done by using elementary properties of norms of matrices and the above results. Let E be the $J \times J$ matrix obtained by compressing $\mathcal{E}$ with one row and column for each level,

$$E_{lk} = \rho(\mathcal{E}_{LK}),$$

where L and K are the index sets for levels l and k. That is, $\mathcal{E}_{LK}$ is the submatrix of $\mathcal{E}$ obtained by extracting only those rows associated with level l and columns associated with level k.

It is obvious that

$$\rho(\mathcal{E}) \leq \rho(E).$$

Since E is symmetric, the maximum row sum equals the maximum column sum, that is, **(F3)**, $||E||_\infty = ||E||_1$, thus

$$\begin{aligned} \rho(E) &\leq \sqrt{||E||_\infty ||E||_1}, \\ &= ||E||_1, \\ &= \max_i \sum_j |E_{ij}|. \end{aligned}$$

Thus we must obtain bounds on E_{ij}. To do this we need the following lemma. □

Lemma 9: *Let n_c denote the maximum number of nonzeros in any column of the matrix A; then the spectral radius of A is bounded by $\sqrt{n_c}$ times the maximal 2-norm of the rows of A.*

$$\rho(A) \le \sqrt{n_c} \max_i \left(\sum_j A_{ij}^2 \right)^{1/2}.$$

PROOF: Recall the definition of

$$\rho(A) = \max_{x \neq 0} \frac{|Ax|_2}{|x|_2}.$$

Let θ_{ij} be 1 if $A_{ij} \neq 0$ and zero otherwise. The use of the Cauchy-Schwarz inequality for sums gives

$$\begin{aligned}
|Ax|_2^2 &= \sum_i \left(\sum_j A_{ij} x_j \right)^2, \\
&= \sum_i \left(\sum_j A_{ij} \theta_{ij} x_j \right)^2, \\
&\le \sum_i \left(\sum_j A_{ij}^2 \right) \left(\sum_j \theta_{ij} x_j^2 \right), \\
&\le \left(\max_i \sum_j A_{ij}^2 \right) \sum_i \sum_j \theta_{ij} x_j^2, \\
&\le \left(\max_i \sum_j A_{ij}^2 \right) \sum_j \left(\sum_i \theta_{ij} \right) x_j^2, \\
&\le n_c \left(\max_i \sum_j A_{ij}^2 \right) |x|_2^2.
\end{aligned}$$

Taking the square root of both sides completes the proof.

Combining the above lemma and (**F1**) and (**F2**) one obtains the bound

$$\begin{aligned}
E_{lk} &= \rho(\mathcal{E}_{LK}), \\
&\le \sqrt{n_c} \max_{i \in \text{level } l} \left(\sum_{j \in \text{level } k} \mathcal{E}_{ij}^2 \right)^{1/2}, \\
&\le \sqrt{n_c} \max_{i \in \text{level } l} \left(\gamma^{-|l-k|} \gamma^{2|l-k|} \right)^{1/2}, \\
&\le C \gamma^{|l-k|/2}.
\end{aligned}$$

Now apply (**F3**)

$$\begin{aligned}
\rho(\mathcal{E}) &\le \rho(E), \\
&\le \max_i \sum_j |E_{ij}|,
\end{aligned}$$

$$\le C \max_i \sum_j \gamma^{|i-j|/2},$$

$$\le \frac{C}{1-\gamma^{1/2}}.$$

Thus $\rho(\mathcal{E})$ may be bounded independently of the number of levels.

In the multilevel diagonal scaling method the one-dimensional subspace problems are solved exactly; hence ω in Assumption 3 is 1. □

Notes and References for Section 5.3.4

- An early account of this type of proof may be found in Dryja and Widlund [Dry91].
- A proof with the less than optimal J^2 growth in the condition number for the BPX algorithm may be found in Bramble, Pasciak, and Xu [Bra90]. For convex domains they do have the optimal lower bound, but have the $O(J)$ growth in the largest eigenvalue.
- Oswald, realizing the connection between the BPX algorithm and results in Besov space theory, provided an optimal bound for the BPX method in [Osw90]. See also [Osw94].
- Independently, Zhang also derived such bounds for the more general multilevel additive Schwarz methods in [Zha92]. His proof, with some simplifications, is what we have given above. Zhang also communicated to us the proof of Lemma 9.
- See also, slightly later, Borneman and Yserentant [Bor93].
- Other discussions of the relationships between traditional multigrid and multilevel methods may be found in Yserentant [Yse93]. See Bramble [Bra93] for an in-depth discussion of these issues.
- Oswald and Zhang have extended many of their results to the biharmonic problems [Osw92a], [Osw91], [Zha94].
- A completely algebraic proof for the hierarchical basis method that uses techniques similar to those above may be found in Elman and Zhang [Elm95].
- The recent book by Rüde [Rüd93] contains a detailed discussion of multilevel splittings and their stability.

5.4 Indefinite and Nonsymmetric Problems

5.4.1 Abstract Theory

This section may be omitted during a first reading. In this section we will expand the abstract theory given in Section 5.2 to non–self-adjoint elliptic problems that give rise to nonsymmetric matrices. We will only consider the nested subspace case; that is, $\boldsymbol{I}_i$ is the identity operator. Let $b(\cdot,\cdot)$ be the nonsymmetric bilinear form that defines our linear system. That is,

$$b(\boldsymbol{u}^*, v) = f(v) \qquad \forall v \in V.$$

We give a concrete example of $b(\cdot,\cdot)$ below in Section 5.4.2.

The convergence of Krylov subspace methods for nonsymmetric systems, for instance, GMRES, may be characterized by the two quantities

$$c_p = \inf_{u \neq 0} \frac{a(\frac{1}{2}(\boldsymbol{T} + \boldsymbol{T}^T)\boldsymbol{u}, \boldsymbol{u})}{a(\boldsymbol{u}, \boldsymbol{u})} = \inf_{u \neq 0} \frac{a(\boldsymbol{Tu}, \boldsymbol{u})}{a(\boldsymbol{u}, \boldsymbol{u})}$$

and

$$C_p = \sup_{u \neq 0} \frac{a(\boldsymbol{Tu}, \boldsymbol{Tu})}{a(\boldsymbol{u}, \boldsymbol{u})}.$$

Here $a(\cdot, \cdot)$ is an auxiliary inner product needed for the proofs; for second order elliptic PDEs it would be the usual H^1 inner product.

We first state the abstract assumptions needed on the subspaces and the bilinear forms. In addition to Assumption 2 for the symmetric positive definite case we need Assumption 4:

Assumption 4: *Assume that for all $\boldsymbol{T}_i$, there exist a constant $0 < \omega < 2$ and parameters $\delta_i \geq 0$, such that $\sum_i \delta_i$ can be made sufficiently small, so that*

$$a(\tfrac{1}{2}(\boldsymbol{T}_i + \boldsymbol{T}_i^T)\boldsymbol{u}, \boldsymbol{u}) = a(\boldsymbol{T}_i\boldsymbol{u}, \boldsymbol{u}) \geq \omega^{-1} a(\boldsymbol{T}_i\boldsymbol{u}, \boldsymbol{T}_i\boldsymbol{u}) - \delta_i a(\boldsymbol{u}, \boldsymbol{u}).$$

This simply provides a lower bound on the smallest eigenvalue of the symmetric part of $\boldsymbol{T}_i$. In the symmetric, positive definite case, the $\delta_i = 0$. The requirement that $\sum_i \delta_i$ is sufficiently small turns out to be a requirement that the nonsymmetric part be a compact perturbation of a symmetric operator. We will see below how the coarse grid spacing affects these parameters δ_i.

We also replace Assumption 1 with the following:

Assumption 1*:

$$\sum_i a\,(\boldsymbol{T}_i\boldsymbol{u}, \boldsymbol{T}_i\boldsymbol{u}) \geq C_0^{-2} a(\boldsymbol{u}, \boldsymbol{u}).$$

In the symmetric, positive definite case, with $a_i(\cdot, \cdot) = a(\cdot, \cdot)$, this is an immediate consequence of Assumption 1 and Lemma 1 since $a(\boldsymbol{T}_i\boldsymbol{u}, \boldsymbol{T}_i\boldsymbol{u}) = a(\boldsymbol{T}_i\boldsymbol{u}, \boldsymbol{u})$.

Lemma 10: *For indefinite and nonsymmetric problems, the abstract additive Schwarz method satisfies*

$$c_p \geq \omega^{-1} C_0^{-2} - \sum_i \delta_i$$

and

$$C_p \leq 2\omega^2 \rho^2(\mathcal{E}) + (\sum_i \delta_i)^2.$$

PROOF: The lower bound follows from Assumptions 4 and 1*.

$$\begin{aligned} a(\boldsymbol{Tu},\boldsymbol{u}) &= \sum_i a(\boldsymbol{T}_i\boldsymbol{u},\boldsymbol{u}), \\ &\geq \omega^{-1}\sum_i a(\boldsymbol{T}_i\boldsymbol{u},\boldsymbol{T}_i\boldsymbol{u}) - \sum_i \delta_i a(\boldsymbol{u},\boldsymbol{u}), \\ &\geq (\omega^{-1}C_0^{-2} - \sum_i \delta_i) a(\boldsymbol{u},\boldsymbol{u}). \end{aligned}$$

The upper bound is more complex. Assumption 2 gives

$$\begin{aligned} a(\boldsymbol{Tu},\boldsymbol{Tu}) &= a(\sum_i \boldsymbol{T}_i\boldsymbol{u}, \sum_j \boldsymbol{T}_j\boldsymbol{u}), \\ &\leq \sum_i\sum_j a(\boldsymbol{T}_i\boldsymbol{u},\boldsymbol{T}_j\boldsymbol{u}), \\ &\leq \rho(\mathcal{E})\sum_i a(\boldsymbol{T}_i\boldsymbol{u},\boldsymbol{T}_i\boldsymbol{u}). \end{aligned} \tag{5.16}$$

Assumption 4 followed by Cauchy-Schwarz and then Assumption 2 gives

$$\begin{aligned} \sum_i a(\boldsymbol{T}_i\boldsymbol{u},\boldsymbol{T}_i\boldsymbol{u}) &\leq \omega\sum_i a(\boldsymbol{T}_i\boldsymbol{u},\boldsymbol{u}) + \omega\sum_i \delta_i a(\boldsymbol{u},\boldsymbol{u}), \\ &\leq \omega a(\sum_i \boldsymbol{T}_i\boldsymbol{u}, \sum_j \boldsymbol{T}_j\boldsymbol{u})^{1/2} a(\boldsymbol{u},\boldsymbol{u})^{1/2} + \omega\sum_i \delta_i a(\boldsymbol{u},\boldsymbol{u}), \\ &\leq \omega\rho^{1/2}(\mathcal{E})[\sum_i a(\boldsymbol{T}_i\boldsymbol{u},\boldsymbol{T}_i\boldsymbol{u})]^{1/2} a(\boldsymbol{u},\boldsymbol{u})^{1/2} + \omega\sum_i \delta_i a(\boldsymbol{u},\boldsymbol{u}). \end{aligned}$$

Let $x^2 = \sum_i a(\boldsymbol{T}_i\boldsymbol{u},\boldsymbol{T}_i\boldsymbol{u})$, $x \geq 0$; then

$$x^2 - \omega\rho^{1/2}(\mathcal{E})||\boldsymbol{u}||_a x - \omega\sum_i \delta_i ||\boldsymbol{u}||_a^2 \leq 0.$$

Therefore

$$\begin{aligned} x &\leq ||\boldsymbol{u}||_a \left[\frac{\omega\rho^{1/2}(\mathcal{E})}{2} + \left(\frac{\omega^2\rho(\mathcal{E})}{4} + \omega\sum_i \delta_i\right)^{1/2}\right], \\ &\leq ||\boldsymbol{u}||_a \left[\frac{\omega\rho^{1/2}(\mathcal{E})}{2} + \frac{\omega\rho^{1/2}(\mathcal{E})}{2}\left(1 + \frac{4\sum_i \delta_i}{\omega\rho(\mathcal{E})}\right)^{1/2}\right], \\ &\leq \left[\omega\rho^{1/2}(\mathcal{E}) + \frac{\sum_i \delta_i}{\rho^{1/2}(\mathcal{E})}\right] ||\boldsymbol{u}||_a. \end{aligned}$$

Here we use the fact that $\sum_i \delta_i$ may be made sufficiently small so that we may replace the square root term with the first two terms of its Taylor series; more precisely we require $\sum_i \delta_i \ll \frac{\omega\rho(\mathcal{E})}{4}$. Combining this with (5.16) gives us the upper bound.

$$\begin{aligned}
a(\boldsymbol{T}\boldsymbol{u}, \boldsymbol{T}\boldsymbol{u}) &\le \rho(\mathcal{E}) \sum_i a(\boldsymbol{T}_i\boldsymbol{u}, \boldsymbol{T}_i\boldsymbol{u}), \\
&\le \rho(\mathcal{E}) \left(\omega\rho^{1/2}(\mathcal{E}) + \frac{\sum_i \delta_i}{\rho^{1/2}(\mathcal{E})}\right)^2 a(\boldsymbol{u}, \boldsymbol{u}), \\
&\le 2\rho(\mathcal{E}) \left(\omega^2\rho(\mathcal{E}) + \frac{(\sum_i \delta_i)^2}{\rho(\mathcal{E})}\right) a(\boldsymbol{u}, \boldsymbol{u}).
\end{aligned}$$

□

Before we provide a result for the multiplicative case, we provide two lemmas which follow directly from the definitions and some careful algebraic manipulations.

Lemma 11: *With* $\boldsymbol{E}_{-1} = I$ *and* $\boldsymbol{E}_i = (I - \boldsymbol{T}_i)\boldsymbol{E}_{i-1} = (I - \boldsymbol{T}_i)\cdots(I - \boldsymbol{T}_0)$,

$$a(\boldsymbol{E}_{i-1}\boldsymbol{u}, \boldsymbol{E}_{i-1}\boldsymbol{u}) \le e^{2\sum_{j=0}^{j=i-1}\delta_j} a(\boldsymbol{u}, \boldsymbol{u}).$$

PROOF:

$$\begin{aligned}
a((I - \boldsymbol{T}_i)\boldsymbol{u}, (I - \boldsymbol{T}_i)\boldsymbol{u}) &= a(\boldsymbol{u}, \boldsymbol{u}) - 2a(\boldsymbol{T}_i\boldsymbol{u}, \boldsymbol{u}) + a(\boldsymbol{T}_i\boldsymbol{u}, \boldsymbol{T}_i\boldsymbol{u}), \\
&\le a(\boldsymbol{u}, \boldsymbol{u}) + \left(1 - \frac{2}{\omega}\right) a(\boldsymbol{T}_i\boldsymbol{u}, \boldsymbol{T}_i\boldsymbol{u}) + 2\delta_i a(\boldsymbol{u}, \boldsymbol{u}), \\
&\le (1 + 2\delta_i) a(\boldsymbol{u}, \boldsymbol{u}), \\
&\le e^{2\delta_i} a(\boldsymbol{u}, \boldsymbol{u}).
\end{aligned}$$

The second step follows from Assumption 4 while the third step follows from the fact that $\omega < 2$ and $a(\cdot, \cdot)$ is positive definite. The lemma follows by applying this result to each term of $\boldsymbol{E}_{i-1}$.

□

Lemma 12:

$$\sum_i a(\boldsymbol{T}_i\boldsymbol{E}_{i-1}\boldsymbol{u}, \boldsymbol{T}_i\boldsymbol{E}_{i-1}\boldsymbol{u}) \ge C[\omega^2\rho^2(\mathcal{E}) + (\sum_i \delta_i)^2 + 1]^{-1} \sum_i a(\boldsymbol{T}_i\boldsymbol{u}, \boldsymbol{T}_i\boldsymbol{u}).$$

PROOF: The proof is elementary, but somewhat involved. It may be found in X. Cai and Widlund [Cai93a].

□

Lemma 13: *For sufficiently small* δ_i *the abstract multiplicative Schwarz method, for indefinite and nonsymmetric problems, satisfies*

$$c_p \ge \frac{C(\frac{2}{\omega} - 1)}{[\omega^2\rho^2(\mathcal{E}) + (\sum \delta_i)^2 + 1]\, C_0^2}$$

and

$$C_p \le C.$$

PROOF: The upper bound follows from Lemma 11 and the definition of $\boldsymbol{BA}$. For the lower bound, we use Assumption 4 to obtain

$$a(\boldsymbol{BAu}, \boldsymbol{u}) = \sum_i a(2\boldsymbol{E}_{i-1}\boldsymbol{u}, \boldsymbol{T}_i\boldsymbol{E}_{i-1}\boldsymbol{u}) - \sum_i a(\boldsymbol{T}_i\boldsymbol{E}_{i-1}\boldsymbol{u}, \boldsymbol{T}_i\boldsymbol{E}_{i-1}\boldsymbol{u}),$$

$$\geq \sum_i \left(\frac{2}{\omega} - 1\right) a(\boldsymbol{T}_i\boldsymbol{E}_{i-1}\boldsymbol{u}, \boldsymbol{T}_i\boldsymbol{E}_{i-1}\boldsymbol{u}) - 2\sum_i \delta_i a(\boldsymbol{E}_{i-1}\boldsymbol{u}, \boldsymbol{E}_{i-1}\boldsymbol{u}).$$

We use the previous two lemmas, Lemma 11 and 12, and Assumption 1* to obtain

$$a(\boldsymbol{BAu}, \boldsymbol{u}) \geq \left(\frac{(\frac{2}{\omega} - 1)}{[\omega^2 \rho^2(\mathcal{E}) + (\sum_i \delta_i)^2 + 1]C_0^2} - 2\sum_i \delta_i e^{2\sum_{j=0}^{i-1}\delta_j} \right) a(\boldsymbol{u}, \boldsymbol{u}).$$

Now if the δ_i can be made sufficiently small, the second term is dominated by the first and the proof is complete. □

5.4.2 Convergence for Two Level Overlapping Schwarz

We now consider the more general non–self-adjoint second order elliptic problem,

$$Lu = f \qquad \text{in } \Omega,$$
$$Bu = g \qquad \text{on } \partial\Omega.$$

The operator L is given by

$$L\boldsymbol{u} = -\sum_{i,j} \frac{\partial}{\partial x_i}\left(a_{ij}(x)\frac{\partial \boldsymbol{u}}{\partial x_j}\right) + \sum_i b_i(x)\frac{\partial \boldsymbol{u}}{\partial x_i} + c(x)\boldsymbol{u}.$$

The matrix $a_{ij}(x)$ is assumed to be symmetric and uniformly positive definite. We can allow general boundary conditions, denoted by the operator B above. To simplify notation, however, we consider only the case of homogeneous boundary conditions, $\boldsymbol{u} = 0$ on $\partial\Omega$. Again we can introduce appropriate test functions and apply a suitable Green's formula to obtain the weak form: find $\boldsymbol{u} \in V^h \subset H_0^1(\Omega)$ such that

$$b(\boldsymbol{u}, \boldsymbol{v}) = \int_\Omega f\boldsymbol{v}\,dx \qquad \forall \boldsymbol{v} \in V^h.$$

But now

$$b(\boldsymbol{u}, \boldsymbol{v}) = \int_\Omega \left(\sum_{i,j} a_{ij}(x) \frac{\partial \boldsymbol{u}}{\partial x_i}\frac{\partial \boldsymbol{v}}{\partial x_j} + \sum_i b_i(x)\frac{\partial \boldsymbol{u}}{\partial x_i}\boldsymbol{v} + c(x)\boldsymbol{uv} \right) dx.$$

Because of the assumptions on $a_{ij}(x)$, the first term in $b(\cdot,\cdot)$ is equivalent to the form $a(\cdot,\cdot)$ introduced above. That is, there exist two positive constants c and C such that

$$c||\boldsymbol{u}||_{H^1}^2 \leq ca(\boldsymbol{u}, \boldsymbol{u}) \leq \int_\Omega \sum_{i,j} a_{ij}(x)\frac{\partial \boldsymbol{u}}{\partial x_i}\frac{\partial \boldsymbol{u}}{\partial x_j}\,dx \leq Ca(\boldsymbol{u}, \boldsymbol{u}) \leq C||\boldsymbol{u}||_{H^1}^2.$$

Let $s(\boldsymbol{u}, \boldsymbol{v})$ denote the skew-symmetric part of $b(\boldsymbol{u}, \boldsymbol{v})$, that is,

$$s(\boldsymbol{u}, \boldsymbol{v}) = \int_{\Omega} \left(\sum_i b_i(x) \frac{\partial \boldsymbol{u}}{\partial x_i} \boldsymbol{v} + \frac{\partial b_i(x) \boldsymbol{u}}{\partial x_i} \boldsymbol{v} \right) dx = b(\boldsymbol{u}, \boldsymbol{v}) - b(\boldsymbol{v}, \boldsymbol{u}).$$

Note that

$$s(\boldsymbol{u}, \boldsymbol{u}) = 0 \quad \forall \boldsymbol{u}. \tag{5.17}$$

Also define

$$c(\boldsymbol{u}, \boldsymbol{v}) = \int_{\Omega} (c(x) - \nabla \cdot b(x)) \boldsymbol{u} \boldsymbol{v} \, dx,$$

then

$$b(\boldsymbol{u}, \boldsymbol{v}) = a(\boldsymbol{u}, \boldsymbol{v}) + s(\boldsymbol{u}, \boldsymbol{v}) + c(\boldsymbol{u}, \boldsymbol{v}).$$

The tools needed to verify Assumptions 1*, 2, and 4 and hence apply the general abstract framework given in the previous section for non-self-adjoint problems are the same as those needed for the analysis of the convergence of the finite element method for non-self-adjoint problems.

I1. There exists a constant, C, such that, for all $\boldsymbol{u}, \boldsymbol{v} \in H_0^1(\Omega)$,

$$|b(\boldsymbol{u}, \boldsymbol{v})| \leq C ||\boldsymbol{u}||_a ||\boldsymbol{v}||_a.$$

I2. The skew-symmetric term may be bounded, for all $\boldsymbol{u}, \boldsymbol{v} \in H_0^1(\Omega)$, by

$$|s(\boldsymbol{u}, \boldsymbol{v})| \leq C ||\boldsymbol{u}||_a ||\boldsymbol{v}||_{L^2(\Omega)}.$$

I3. Gårding's inequality: there exists a positive constant C such that for all $\boldsymbol{u} \in H_0^1(\Omega)$,

$$||\boldsymbol{u}||_a^2 - C ||\boldsymbol{u}||_{L^2(\Omega)}^2 \leq b(\boldsymbol{u}, \boldsymbol{u}).$$

I4. The solution of the adjoint equation

$$b(\phi, \boldsymbol{w}) = f(\phi) \qquad \forall \phi \in H_0^1(\Omega)$$

satisfies

$$||\boldsymbol{w}||_{H^{1+\rho}} \leq C ||f||_{L^2(\Omega)},$$

where $\rho \geq 1/2$.

Using these assumptions, it is possible to prove the fundamental lemma for finite element analysis of non-self-adjoint elliptic PDEs, which is due to Schatz [Sch74].

Lemma 14: *For H sufficiently small, that is, $H < H_0$, for some H_0,*

$$||\boldsymbol{T}_0\boldsymbol{u}||_a \leq C||\boldsymbol{u}||_a$$

and

$$||\boldsymbol{T}_0\boldsymbol{u} - \boldsymbol{u}||_{L^2(\Omega)} \leq CH^{\rho}||\boldsymbol{u}||_a.$$

This lemma says essentially that if the coarse grid is fine enough, then one obtains H^1-stability and L^2-approximation results similar to those obtained in the self-adjoint, uniformly elliptic case.

We will need a similar type of bound for $\boldsymbol{T}_i$, $i > 0$.

Lemma 15: *For $i > 0$.*

$$||\boldsymbol{T}_i\boldsymbol{u}||_{L^2(\Omega)} \leq CH||\boldsymbol{T}_i\boldsymbol{u}||_a.$$

This follows because the support of $T_i\boldsymbol{u}$, $i > 0$ lies in a region of diameter H.

Theorem 7: *For the overlapping Schwarz method, using exact solvers on the subproblems, if the overlap is uniformly of diameter $O(H)$ then c_p and C_p can be bounded independently of h and H if H is sufficiently small.*

SKETCH OF PROOF: The proof involves a large number of careful estimates. All details are not given below. We only consider the case of exact projections where $b_i(\cdot, \cdot) = b(\cdot, \cdot)$.

We first derive a bound for Assumption 4. To do this we will need the following estimates.

Lemma 16:

$$|s(\boldsymbol{u} - \boldsymbol{T}_i\boldsymbol{u}, \boldsymbol{T}_i\boldsymbol{u})| \leq CH^{\sigma_i}[a(\boldsymbol{u}, \boldsymbol{u}) + a(\boldsymbol{T}_i\boldsymbol{u}, \boldsymbol{T}_i\boldsymbol{u})]$$

and

$$|c(\boldsymbol{u} - \boldsymbol{T}_i\boldsymbol{u}, \boldsymbol{T}_i\boldsymbol{u})| \leq CH^{\sigma_i}[a(\boldsymbol{u}, \boldsymbol{u}) + a(\boldsymbol{T}_i\boldsymbol{u}, \boldsymbol{T}_i\boldsymbol{u})]$$

where $\sigma_i = 1$ for $i > 0$ and $\sigma_i = \rho$ for the coarse grid space.

PROOF OF LEMMA 16: For $i = 0$, applying inequality (**I2**) one obtains

$$\begin{aligned}|s(\boldsymbol{u} - \boldsymbol{T}_i\boldsymbol{u}, \boldsymbol{T}_i\boldsymbol{u})| &\leq C||\boldsymbol{T}_i\boldsymbol{u}||_a||\boldsymbol{u} - \boldsymbol{T}_i\boldsymbol{u}||_{L^2(\Omega)},\\ &\leq CH^{\rho_i}||\boldsymbol{T}_i\boldsymbol{u}||_a||\boldsymbol{u}||_a,\\ &\leq CH^{\rho_i}(||\boldsymbol{T}_i\boldsymbol{u}||_a^2 + ||\boldsymbol{u}||_a^2).\end{aligned}$$

The second line follows from Lemma 14.

For $i > 0$, since $s(\boldsymbol{T}_i\boldsymbol{u}, \boldsymbol{T}_i\boldsymbol{u}) = 0$, by (5.17)

$$\begin{aligned}|s(\boldsymbol{u} - \boldsymbol{T}_i\boldsymbol{u}, \boldsymbol{T}_i\boldsymbol{u})| &= |s(\boldsymbol{u}, \boldsymbol{T}_i\boldsymbol{u})|,\\ &\le C||\boldsymbol{u}||_a||\boldsymbol{T}_i\boldsymbol{u}||_{L^2(\Omega)},\\ &\le CH||\boldsymbol{u}||_a||\boldsymbol{T}_i\boldsymbol{u}||_a,\\ &\le CH(||\boldsymbol{u}||_a^2 + ||\boldsymbol{T}_i\boldsymbol{u}||_a^2).\end{aligned}$$

This follows Inequality (**I2**) and Lemma 15.

The bound for $c(\cdot, \cdot)$ is obtained by using the Cauchy-Schwarz inequality and Lemma 14 or Lemma 15. For $i = 0$,

$$\begin{aligned}|c(\boldsymbol{u} - \boldsymbol{T}_0\boldsymbol{u}, \boldsymbol{T}_0\boldsymbol{u})| &\le C||\boldsymbol{u} - \boldsymbol{T}_0\boldsymbol{u}||_{L^2(\Omega)}||\boldsymbol{T}_0\boldsymbol{u}||_{L^2(\Omega)},\\ &\le CH^\rho||\boldsymbol{u}||_a||\boldsymbol{T}_0\boldsymbol{u}||_a,\\ &\le CH^\rho(||\boldsymbol{u}||_a^2 + ||\boldsymbol{T}_0\boldsymbol{u}||_a^2).\end{aligned}$$

For $i > 0$,

$$\begin{aligned}|c(\boldsymbol{u} - \boldsymbol{T}_i\boldsymbol{u}, \boldsymbol{T}_i\boldsymbol{u})| &\le |c(\boldsymbol{u}, \boldsymbol{T}_i\boldsymbol{u})| + |c(\boldsymbol{T}_i\boldsymbol{u}, \boldsymbol{T}_i\boldsymbol{u})|,\\ &\le C||\boldsymbol{u}||_{L^2(\Omega)}||\boldsymbol{T}_i\boldsymbol{u}||_{L^2(\Omega)} + C||\boldsymbol{T}_i\boldsymbol{u}||_{L^2(\Omega)}^2,\\ &\le CH||\boldsymbol{u}||_a||\boldsymbol{T}_i\boldsymbol{u}||_a + CH^2||\boldsymbol{T}_i\boldsymbol{u}||_a^2,\\ &\le CH(||\boldsymbol{u}||_a^2 + ||\boldsymbol{T}_i\boldsymbol{u}||_a^2).\end{aligned}$$

From the definition of $\boldsymbol{T}_i$,

$$b(\boldsymbol{T}_i\boldsymbol{u}, \boldsymbol{T}_i\boldsymbol{u}) = b(\boldsymbol{u}, \boldsymbol{T}_i\boldsymbol{u}).$$

Thus

$$\begin{aligned}0 &= b(\boldsymbol{T}_i\boldsymbol{u} - \boldsymbol{u}, \boldsymbol{T}_i\boldsymbol{u}),\\ &= a(\boldsymbol{T}_i\boldsymbol{u} - \boldsymbol{u}, \boldsymbol{T}_i\boldsymbol{u}) + s(\boldsymbol{T}_i\boldsymbol{u} - \boldsymbol{u}, \boldsymbol{T}_i\boldsymbol{u}) + c(\boldsymbol{T}_i\boldsymbol{u} - \boldsymbol{u}, \boldsymbol{T}_i\boldsymbol{u}),\end{aligned}$$

or

$$\begin{aligned}a(\boldsymbol{T}_i\boldsymbol{u}, \boldsymbol{u}) &= a(\boldsymbol{T}_i\boldsymbol{u}, \boldsymbol{T}_i\boldsymbol{u}) + s(\boldsymbol{T}_i\boldsymbol{u} - \boldsymbol{u}, \boldsymbol{T}_i\boldsymbol{u}) + c(\boldsymbol{T}_i\boldsymbol{u} - \boldsymbol{u}, \boldsymbol{T}_i\boldsymbol{u}),\\ &\ge a(\boldsymbol{T}_i\boldsymbol{u}, \boldsymbol{T}_i\boldsymbol{u}) - |s(\boldsymbol{T}_i\boldsymbol{u} - \boldsymbol{u}, \boldsymbol{T}_i\boldsymbol{u})| - |c(\boldsymbol{T}_i\boldsymbol{u} - \boldsymbol{u}, \boldsymbol{T}_i\boldsymbol{u})|,\\ &\ge a(\boldsymbol{T}_i\boldsymbol{u}, \boldsymbol{T}_i\boldsymbol{u}) - CH^{\sigma_i}(a(\boldsymbol{u}, \boldsymbol{u}) + a(\boldsymbol{T}_i\boldsymbol{u}, \boldsymbol{T}_i\boldsymbol{u})),\\ &\ge (1 - CH^{\sigma_i})a(\boldsymbol{T}_i\boldsymbol{u}, \boldsymbol{T}_i\boldsymbol{u}) - CH^{\sigma_i}a(\boldsymbol{u}, \boldsymbol{u}).\end{aligned}$$

Thus if H is small enough so that $(1 - CH^{\sigma_i}) > 0$, then

$$\omega^{-1} = \max_i(1 - CH^{\sigma_i})$$

and $\delta_i = CH^{\sigma_i}$.

Since the subspaces used here are identical to those used in Theorem 1, Assumption 2 can be handled exactly as in that proof. This is because Assumption 2 is only concerned with the orthogonality of the subspaces, not the bilinear forms.

We must also verify Assumption 1*. We use the same decomposition of $\boldsymbol{u}$ that is used in the proof of Theorem 1. Note that this decomposition is independent of the particular bilinear form $b(\cdot,\cdot)$. The proof requires only Inequality (**I1**), the definition of $\boldsymbol{T}_i$, the H^1-stability property of the $\boldsymbol{u}_i$ and Lemma 14.

We begin by obtaining an upper bound on $b(\cdot,\cdot)$.

$$\begin{aligned} b(\boldsymbol{u},\boldsymbol{u}) &= \sum_{i=0}^{p} b(\boldsymbol{u},\boldsymbol{u}_i), \\ &\leq \sum_{i} b(\boldsymbol{T}_i\boldsymbol{u},\boldsymbol{u}_i), \\ &\leq C\sum_{i} ||\boldsymbol{T}_i\boldsymbol{u}||_a||\boldsymbol{u}_i||_a, \\ &\leq C||\boldsymbol{u}||_a(\sum_{i} ||\boldsymbol{T}_i\boldsymbol{u}||_a^2)^{1/2}. \end{aligned}$$

We next obtain a lower bound on $b(\cdot,\cdot)$.

$$\begin{aligned} b(\boldsymbol{u},\boldsymbol{u}) &\geq ||\boldsymbol{u}||_a^2 - C||\boldsymbol{u}||_{L^2(\Omega)}^2, \\ &\geq ||\boldsymbol{u}||_a^2 - C||\boldsymbol{u} - \boldsymbol{T}_0\boldsymbol{u}||_{L^2(\Omega)}^2 - C||\boldsymbol{T}_0\boldsymbol{u}||_{L^2(\Omega)}^2, \\ &\geq ||\boldsymbol{u}||_a^2 - CH^{2\rho}||\boldsymbol{u}||_a^2 - C||\boldsymbol{u}||_a||\boldsymbol{T}_0\boldsymbol{u}||_a, \\ &\geq (1 - CH^{2\rho})||\boldsymbol{u}||_a^2 - C||\boldsymbol{u}||_a||\boldsymbol{T}_0\boldsymbol{u}||_a. \end{aligned}$$

Now combine the two inequalities and cancel the common factor $||\boldsymbol{u}||_a$. Move the term $||\boldsymbol{T}_0\boldsymbol{u}||_a$ to the other side, then square both sides to obtain the required bound

$$\sum_{i} a(\boldsymbol{T}_i\boldsymbol{u},\boldsymbol{T}_i\boldsymbol{u}) \geq Ca(\boldsymbol{u},\boldsymbol{u}). \qquad \square$$

Notes and References for Section 5.4.2

- The convergence theory for the non–self-adjoint case is taken from X. Cai and Widlund [Cai92c], [Cai93a]. See also X. Cai [Cai90], [Cai91], [Cai94].
- X. Cai and Widlund also consider the analysis of certain substructuring and multilevel methods.
- Analysis of a particular iterative substructuring method for non-self-adjoint problems may be found in X. Cai, Gropp, and Keyes [Cai92b].
- The assumptions that we have made on the operator L make the lower order terms a relatively compact perturbation of the second order term. In fact, similar analysis, in certain cases, may be carried out when these assumptions are not satisfied, see Bramble, Leyk, and Pasciak [Bra93a].

A1
Preconditioners and Accelerators

IN THIS APPENDIX, we briefly introduce the concepts of preconditioners and accelerators for the solution of linear systems of equations.

Consider the nonsingular linear system

$$Au^* = f.$$

Iterative methods are techniques for improving the accuracy of an approximate solution. Let u be an approximate solution, and denote the error by $e = u^* - u$; then

$$Ae = f - Au = r.$$

Now assume we have a linear algorithm for the approximate solution of $Ae = r$. Using our algorithm B we can find an approximation to the error

$$e \approx Br.$$

Thus the new, corrected approximate solution is given by

$$u^{\text{new}} \leftarrow u + Br.$$

Note that though we write our approximate solver as B it is not necessary that we have a matrix representation of either B or B^{-1}. In general, the effect of B is merely calculated by a (perhaps very complicated) computer code. The matrix operator B (or sometimes B^{-1}) is referred to as the **preconditioner.** The decomposition of A into $A = B^{-1} + (A - B^{-1})$ is often referred to as a **splitting** of A.

In developing preconditioners, the goal is to construct algorithms B that are inexpensive to apply (in terms of floating point operations and/or interprocessor communication) but provide fast convergence, that is, require a small number of iterations to achieve an accurate solution.

A1.1 Simple Preconditioners

In this section, we present a few simple, standard preconditioners. These methods are not particularly powerful, but they are easily understood and implemented.

Example P1: Jacobi Preconditioner

In the Jacobi method the approximate solver is simply $B = D^{-1}$, the inverse of the diagonal of the matrix A. Then

$$u^{\text{new}} \leftarrow u + D^{-1}(f - Au),$$

equivalently,

$$u^{\text{new}} \leftarrow D^{-1}(f - (A - D)u),$$

or, componentwise,

$$u_j^{\text{new}} \leftarrow \frac{1}{A_{jj}} \left(f_j - \sum_{k \neq j} A_{jk} u_k \right).$$

An obvious drawback of this approach is that the new u_j, $j > 1$, is not calculated by using the newest values of u_k, $k < j$. The Gauss-Seidel method addresses this by using the most recently computed values of u in the updating process.

Example P2: Gauss-Seidel Preconditioner

$$u_j^{\text{new}} \leftarrow \frac{1}{A_{jj}} \left(f_j - \sum_{k<j} A_{jk} u_k^{\text{new}} + \sum_{k>j} A_{jk} u_k \right).$$

The matrix representation of the preconditioner B can be written explicitly for the Gauss-Seidel method (note that one never needs to form it explicitly). If we express A as $A = L + D + U$, where L is the lower triangular part of A, D the diagonal, and U the upper triangular part, then $B = (L + D)^{-1}$. That is, we are approximating the inverse of A by solving a triangular system consisting of the lower triangular part of A. Using this notation

$$u^{\text{new}} \leftarrow u + (L + D)^{-1}(f - Au).$$

Example P3: Block Jacobi and Gauss-Seidel Preconditioners

A natural extension of Jacobi and Gauss-Seidel is to partition the unknowns into blocks and update all the coefficients in a block simultaneously by solving a small linear system associated with just those unknowns. Rather than solving the subproblems exactly, it may make sense to solve them iteratively to some tolerance. Such a procedure is called an **inner iteration**. The convergence of the **outer iteration** will depend in a complicated way on the tolerances, number of iterations, and quality of the preconditioner used in the inner iterations.

Example P4: Incomplete Factorization Preconditioner

If we had an LU factorization of the matrix A, then using $U^{-1}L^{-1}$ as a preconditioner would result in a method which converges to machine precision in one iteration. In general, for sparse matrices, the factors L and U are much denser than A and it takes many floating point operations to form them explicitly. One could, however, approximate L and U by using an incomplete factorization. One technique is to apply the standard LU factorization, but dropping all elements smaller than a given tolerance. This can reduce the memory needed to store the factorization dramatically and also decrease the computation time.

The simplified linear equation solvers (SLES) component of PETSc (see Appendix 2) contains all of the above preconditioners as well as several others.

A1.2 Accelerators: Krylov Subspace Methods

In this section we introduce some methods for accelerating the convergence of the basic iteration.

Example A1: Richardson's Method

If u^n denotes the approximate solution after n iterations of the solution process, we can write the iteration as

$$u^{n+1} \leftarrow u^n + \tau B(f - Au^n).$$

This scheme is often called Richardson's method. The scalar τ is an extra parameter called a damping factor, which may be chosen to improve convergence. We now derive a convergence estimate for Richardson's method and calculate the optimal τ.

$$\begin{aligned} u^n - u^* &= u^{n-1} - u^* + \tau B(Au^* - Au^{n-1}), \\ &= (I - \tau BA)(u^{n-1} - u^*). \end{aligned}$$

This implies

$$||e^n||_2 \leq \max\{|\lambda^n_{\min}(I - \tau BA)|, |\lambda^n_{\max}(I - \tau BA)|\}||e^0||_2.$$

The minimum value of $\max\{|\lambda_{\min}(I - \tau BA)|, |\lambda_{\max}(I - \tau BA)|\}$ is obtained when

$$\tau = 2/[\lambda_{\max}(BA) + \lambda_{\min}(BA)].$$

If we let

$$\kappa(BA) = \frac{\lambda_{\max}(BA)}{\lambda_{\min}(BA)}$$

denote the **condition number** of BA then

$$||e^n||_2 \leq \left[\frac{\kappa(BA) - 1}{\kappa(BA) + 1}\right]^n ||e^0||_2.$$

We see that the convergence rate depends on the condition number, $\kappa(BA)$, of BA. The smaller $\kappa(BA)$ the faster the convergence. Unfortunately, in order to use the optimal value of τ one needs knowledge of the extreme eigenvalues of BA, which is normally not available.

Note that u^{n+1} is obtained by using only information about u^n, and none of the previous u^m, $m < n$ is used explicitly. We will refer to all techniques that use information about more than one previous iteration as accelerator techniques. These include the Chebychev method, the preconditioned conjugate gradient method (PCG), GMRES, and BiCG-stab. For many preconditioners B the Richardson iteration (say with $\tau = 1$) may not even converge. When an accelerator is used with that same preconditioner, often very fast convergence may be obtained. For these reasons, simple Richardson iteration is almost never used in practice.

Example A2: Chebychev

The Chebychev method is a method of accelerating the convergence rate of Richardson's iteration by taking linear combinations of the iterates obtained there. The Chebychev method is a particular instance of the general class of **Krylov subspace methods.** Krylov methods consist of first generating a Krylov subspace given by

$$K_n(r, A, B) = \text{span}\{Br, (BA)Br, (BA)^2 Br, \ldots, (BA)^{n-1} Br\}.$$

An approximate solution to the linear system is then obtained by minimizing the error, in various norms, over all potential solutions in $K_n(r, A, B)$. For the Chebychev method this can be done by using a three term recurrence relation directly related to the three term recurrence relation for Chebychev polynomials.

The Chebychev method has several problems. First, it requires good estimates for the parameter τ and a bound such that $-\alpha \le \lambda(I - \tau BA) \le \alpha$. Second, and more fundamentally, the minimization is taken over all $-\alpha \le \lambda \le \alpha$, rather than just λ_i, the eigenvalues of $I - \tau BA$. The third problem is that above we implicitly assumed that the eigenvalues λ_i are all real. In general, when the eigenvalues are complex, the exact solution of the minimization problem is unknown.

Example A3: The Conjugate Gradient Method

Another Krylov subspace method is the conjugate gradient method. This method is generally only suitable for symmetric positive definite matrices. As with the Chebychev method, it uses a three term recurrence relation. This means that only a small number of work vectors need be kept during the iteration. Unlike the Chebychev method it requires no user supplied parameters. In fact, the method can be shown to be optimal for the symmetric, positive definite class of problems in the sense that it minimizes the residual (in a suitable norm) over a Krylov subspace where the dimension increases by one in every iteration. If our preconditioned operator BA has a low condition number or the eigenvalues are clustered, then the convergence will be very rapid. We strongly recommend the conjugate gradient method of choice for all symmetric, positive definite problems.

Example A4: GMRES

With the GMRES method all of the previous iterations must be kept for the calculation of the next iteration. When a large number of iterations are needed this can increase storage requirements greatly. To prevent this, it is possible to use **restarted GMRES**. This means that after a certain number of iterations, say 10 or 20, the iteration is restarted, by using the most recent approximation as the initial guess. Since with the algorithms presented in this book rarely more than 10 or 20 iterations are needed, we often do not need restarts.

Different Krylov subspace methods are suitable for different problems and situations. Our favorites are the preconditioned conjugate gradient (PCG) method for symmetric positive definite matrices and GMRES or BiCG-stab for all others. Since PCG and GMRES outperform Chebychev, we do not recommend its use in practice.

Most variants of the Krylov subspace methods perform very similarly for well-conditioned problems. Since the goal of this book is to describe a class of very good preconditioners, many of the extra features of particular Krylov subspace methods are unimportant for the material in the book. An understanding of the subtleties of the Krylov subspace methods is not needed for any of the material in this book. The important point is that convergence of Krylov subspace methods depends on the distribution of the **eigenvalues** or the **pseudospectrum** of the preconditioned operator BA. The more tightly clustered the spectrum the faster the convergence. For symmetric positive definite problems, to a first approximation, the clustering may be measured by the condition number of BA, which is simply the ratio of the largest eigenvalue to the smallest. Most proofs for domain decomposition algorithms involve calculating bounds on the extreme eigenvalues.

The Krylov subspace package component of PETSc (see Appendix 2) has a library of routines which apply several different standard accelerators, including BiCG-stab, CG-squared, and two versions of transpose free QMR. All of the Krylov subspace routines are accessed in exactly the same manner by the application programmer and all are suitable for use on parallel computers. Once an application code runs with, for instance, GMRES, it will run without change using any of the above methods.

It is extremely important to note that the preconditioner B and the accelerator are completely independent. In general, any preconditioner can be used with any accelerator, except for the PCG method, which requires a symmetric preconditioner and operator. It is also important to note that when A is nonsingular the preconditioner must be of full rank. If B has a nontrivial null space, the component of r in that space would never be corrected.

The convergence of iterative methods for the solution of linear systems depends strongly on the distribution of the eigenvalues or singular values of the matrix representation of the finite dimensional linear operator. The matrix that is used can be easily obtained from the particular application problem by, for instance, using the usual nodal basis with the finite element method. Other representations of the finite dimensional linear operator may be much better conditioned. One can view preconditioning then as merely a change of basis. In some algorithms the change of basis has a direct physical meaning, in others such an interpretation may be of little use.

It is possible, though potentially dangerous, to apply the accelerator methods within inner iterations. One can, for instance, run GMRES with a preconditioner B, and as part of the application of the preconditioner use GMRES as a solver. Convergence proofs for this type of method are generally difficult to construct. However, in practice, if the inner iterations are solved accurately enough, these methods can perform well.

Notes and References for Section A1

- Three recent texts on iterative methods are by Axelsson [Axe94], Hackbusch [Hac94], and Saad [Saa95].
- Discussions of the inner-outer iteration problem and an analysis of the Chebychev and second-order Richardson methods when each step of the iteration is carried out inexactly are given by Golub and Overton [Golu88]. A similar analysis of inexact, preconditioned steepest descent has been carried out by Munthe-Kaas [Mun87]. A similar understanding of Krylov subspace methods seems much harder and there is no complete analysis along these lines.
- For a recent survey paper on Krylov subspace methods see Freund, Golub, and Nachtigal [Fre92].
- Recently, Saad [Saa93] observed that if one keeps an extra vector for each step of the GMRES method, one is free to change the preconditioner at each iteration. This would allow one to use Krylov subspace methods freely as part of the preconditioner. Saad refers to this algorithm as flexible GMRES (FGMRES).

A2

Software for Numerical Parallel Computing

IMPLEMENTING DOMAIN DECOMPOSITION algorithms can be challenging because of the number of separate components that are involved. For example, a simple implementation of an overlapping Schwarz method must provide:

- matrices for each overlapping region,
- a linear solver for each region,
- a matrix for a coarse grid problem (for multilevel methods),
- matrices or other representations for the transfer of data between the coarse grid and subdomain problems, and
- a Krylov subspace method.

For some simple problems on regular domains it is relatively easy to assemble these components, but as the problem becomes more complex (general domain, matrices, transfer operators, distributed-memory vectors, etc.), it becomes increasingly difficult to juggle all of the pieces. This appendix describes an approach for *designing* software for domain decomposition algorithms that simplifies the process of writing domain decomposition programs and a package written around these ideas. This approach centers around choosing the appropriate mental model or *abstraction* and understanding how to implement that abstraction in a programming language. This will seem to take us far afield from domain decomposition and from parallel computing, but will in fact give us all of the necessary tools to write domain decomposition software.

We must emphasize that all of the software described in this appendix is *experimental* and *subject to change*.

A2.1 Abstraction in Programming

In old programming languages such as Fortran 77, there are a limited number of data types built into the language. For Fortran 77, these types are integer, character, and floating point numbers and dense arrays of integers and floating point numbers. The language provides no mechanism that allows the programmer to construct additional data types that may be appropriate for her or his particular application. This means that all data must be stored and accessed directly in dense arrays. Thus computer codes written in Fortran 77 which involve higher level objects such as sparse matrices and unstructured grids become complicated and difficult to understand and modify. The programmer must constantly worry about the details and cannot easily obtain a broad overview of the implementation. The codes also become very dependent on the storage formats chosen for these abstract objects.

Other languages, such as Fortran 90, Pascal, Ada, C and C++, also have a limited number of basic data types, but the languages have additional mechanisms that allow the user to construct new data structures by combining previously defined ones. These are called structures in C, classes in C++ and derived data types in Fortran 90. These user-defined data structures can often be treated by the programmer much like the standard data types. This allows many of the details of code to be hidden from the application programmer, so that he or she may program directly with higher level abstractions like sparse matrices, grids, discretizations, and vectors without worrying about the details of the implementation.

The examples discussed in the book, as well as the accompanying software, are written in C. Certain modules are written in Fortran 77; however, these are not directly accessed by the application programmer. The software discussed here is a subset of Portable, Extensible Toolkit for Scientific computing (PETSc), written by William Gropp and Barry Smith (later joined by Lois Curfman McInnes). The PETSc package is a large suite of data structures and routines which are intended as building blocks for the construction of portable, parallel, efficient application programs for the solution of problems in scientific computation, especially the solution of partial differential equations. All of its features are directly usable from Fortran 77 or C, though, because of the extra flexibility of the C language, the C interface is somewhat cleaner.

The PETSc package is predicated on two main principles: **data hiding** (also called **data encapsulation**, an important aspect of object-oriented programming) and **software layering**. In data hiding the particular representation of the data is hidden from all code except the few routines that must directly manipulate the data. We refer to these functions as **primitives** for that data type. For instance, a parallel conjugate gradient routine need not know the format in which the vectors and linear operators are stored; only the vector and matrix manipulation routines need to know those details. In software layering, higher level data structures and routines are built by using lower level routines and data structures, which may be built from even simpler data structures, and so on; see Figure A2.1. For instance, software for a nonlinear equation solver could be built on top of a package for the solution of linear systems which is built on software for the manipulation of sparse matrices. Moreover, the routines and data structures on each level are not directly aware of the details of the implementation on the other levels. In this way, for instance, the sparse matrix routines could be completely rewritten without any need for changing the nonlinear equation solver code.

One of the advantages of data hiding and software layering is that it allows several (or many) programmers to develop a large application code without requiring each programmer to understand both the entire application and the underlying architectural details of the machine for which the code is being developed. Instead, different programmers can focus on different aspects of the programming task. For instance, one programmer may develop the code for interacting with the operating system and special

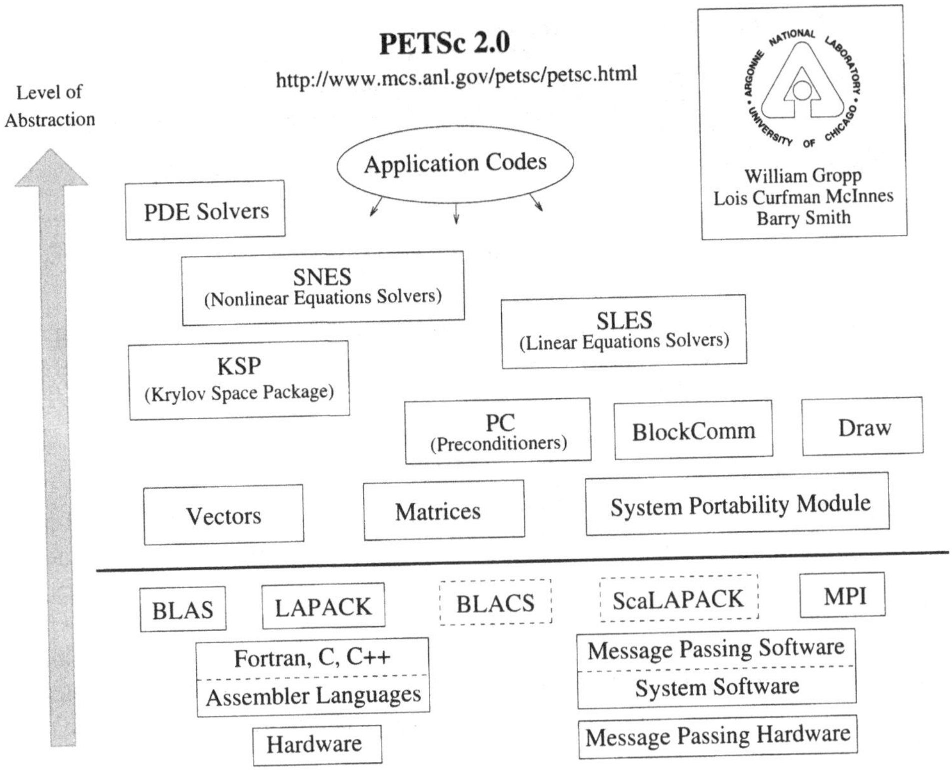

Figure A2.1. The PETSc Package

purpose hardware. A second programmer may focus on the code which performs the numerical computation. Other programmers would concentrate on the user interface, the graphics, the interface related to the particular application, and so forth. Many of the libraries developed in this way can be immediately available for other projects. PETSc is intended both as a prototype of such libraries and as the base for application codes.

We adopt the term **application programmer** to refer to someone, often not a professional programmer, who writes an application code by combining various library routines. Such a person generally understands the application problem in great detail and is not interested in machine level details. A **library developer** is often a professional programmer, or a numerical analyst, who is not an expert at the particular application area but has training in programming, numerical analysis, machine architectures, and so on. The library developer will often work with an application programmer developing the libraries and data structures which are needed. The library developer, however, is obligated to choose data structures and routine interfaces to make her or his code as flexible and reusable as reasonable. In particular, the library developer *must not* foist particular data structures onto the application programmer. On small projects it is possible for the application programmer and library developer to be the same person, especially if a package of library routines such as PETSc is already available.

```
SUBROUTINE CG(M,DESCRA,AR,IA1,IA2,INFORM,DESCRL,LR,IL1,IL2,DESCRU,
*             UR,IU1,IU2,DESCRAN,ARN,IAN1,IAN2,DESCRLN,LRN,ILN1,
*             ILN2,DESCRUN,URN,IUN1,IUN2,VDIAG,B,X,EPS,ITMAX,
*             ERR,ITER,IERROR,Q,R,S,W,P,PT1,IAUX,LIAUX,AUX,LAUX)
```

Figure A2.2. Calling Sequence for a Conjugate Gradient Routine in Fortran 77

A2.2 Abstract Data Types

In this section we introduce several data types which are used extensively by the higher level routines in PETSc. These include vectors, sparse matrices and grids. These are all usable with either Fortran 77 or C, though we recommend the use of C.

A2.2.1 Context Variables

Routines which operate on complicated data structures often have many optional parameters. In Fortran 77, this is dealt with by either having very long argument sequences for the routines or using common blocks. The problem with the former approach is horrible code; with the latter nested calls to the routines become impossible because the action of the routine now depends on the global variables (the variables in the common blocks), not just the arguments to the routines. Common blocks also make it difficult to call libraries written in Fortran 77 from other languages. In Figure A2.2 we give the calling sequence for a conjugate gradient routine designed by using a proposed sparse BLAS standard written in Fortran 77. We consider code like this virtually unusable.

To deal with this problem, we have chosen to use context variables. In their simplest form, context variables are simply a way of avoiding long calling sequences. More generally, they can be used to avoid all global variables and thus allow the flexible nested use of routines. The standard use of context variables is the following:

- create context variable,
- set various properties of the context variable,
- use the context variable,
- set various properties of the context variable,
- use the context variable,
- ...
- free any space used by the context variables.

To increase flexibility even more, the context variables themselves generally have pointers to functions which act on the data. In this way, the application programmer can, for instance, change the vector operations that are used in the iterative solvers at run time. We will demonstrate this use of context variables in the next section.

A2.2.2 Vectors

We begin with vectors, since they are perhaps the simplest to understand, and yet still have many of the important features of the other data types.

```
typedef struct {
  void *(*create_vector)(),    /* Routine returns a single vector */
        (*destroy_vector)(),   /* Free a single vector */
        (*dot)(),                 /* z = x^H * y */
        (*norm)(),                /* z = sqrt(x^H * x) */
        (*max)(),                 /* z = max(|x|);  */
        (*axpy)(),                /*  y = y + alpha * x */
  ...
  void *vecP;                  /* private context */
  void (*vecPDestroy)();
} Vec;
```

Figure A2.3. The Vector Object

In standard Fortran 77 programming, a vector is simply a one dimensional array of numbers stored in consecutive memory locations. On parallel computers there are many potential storage patterns. Even on sequential machines the data layout of the application may require noncontiguous storage of the vector elements, for instance, for an octtree grid. To a mathematician, a vector is an abstract object upon which certain operations can be performed, specifically addition of vectors and multiplication by scalars. No explicit representation of vectors is required. We adopt the mathematician's point of view to allow us the flexibility needed for a host of applications.

To implement the "representation independent vectors," we use vector objects. In Figure A2.3, we give part of the C structure which defines our vector object. The vector object contains two parts, a list of all the vector primitives for this particular vector type, and a private vector context that contains additional information needed by the particular implementation. On a serial machine this may simply be a pointer to an integer which contains the size of the vector. On a parallel machine, it may point to a C structure which contains both the length of the piece which resides on that particular processor and a list of processors sharing the vector. For an out-of-core vector implementation, it may be even more complicated. In C++ the vector operations could be implemented as a **class.** The use of C, on the other hand, makes it possible to provide full support for Fortran 77 programmers who desire to continue to code their application in Fortran 77.

Not only does the vector object contain routines for performing BLAS type operations on vectors; it also contains routines for generating more vectors. Many of the Krylov subspace methods, for instance, GMRES, require a number of work vectors and that number is only known at run time. The GMRES routine then uses `(VecDuplicate)()` to obtain the work vectors needed and `(VecDestroy)()` when it is done with the work space.

All routines that use vectors operate on vectors independently of the underlying storage format of the vector. This is especially important for parallel computers.

Of course, for any storage format chosen for the vectors, one must write the required routines. The advantage is that once they are written and debugged, they can be used immediately in any application which uses the vector context.

PETSc has both serial and parallel implementations of the vector routines. Others can be written as needed.

A2.2.3 Krylov Subspace Methods

PETSc has a suite of Krylov subspace methods for the iterative solution of linear systems, called KSP. These are all programmed with common calling sequences and use the abstract vector types as described in the previous section. Because of its data-structure-neutral implementation, the KSP software is immediately usable on both parallel and sequential machines. Application programmers are free to use whatever data structures are most appropriate for their representation of the matrix operator, their preconditioners as well as their vectors. Several defaults, which are believed to be suitable for many applications, are included.

A2.2.4 Sparse Matrices

The sparse matrices use a form of context variables of type `Mat`. A number of different sparse matrix storage formats are supported, including ones that dynamically allocate memory as required. For users of the sparse matrices, the details of the sparse matrix storage pattern are unimportant.

To keep the data storage of the sparse matrices hidden, we access the sparse matrices only through a small set of well-defined routines, the matrix primitives. These routines allow us to add values to matrices, perform multiplications by matrices, factor matrices, eliminate rows and columns of matrices, and so on.

A2.2.5 Complete Linear System Solvers

PETSc is a hierarchical set of libraries that allows the application programmer to choose how abstract (and hence simpler to use) an interface to the software he or she would like to use. The Simplified Linear Equation Solver (SLES) provides a simple consistent interface for the solution of linear systems. Both direct methods and a variety of iterative methods are available, all with exactly the same interface to the user's application code. SLES is built upon and uses the code from both the sparse matrix libraries and KSP, hence in itself is a rather small package. It, however, saves the less technically oriented users from having to provide any of their own data structures or routines.

A2.2.6 Nonlinear Solvers

PETSc also has data structural neutral software for the solution of nonlinear systems of equations (SNES). The basic approach is Newton and truncated (approximate) Newton techniques. The code provides an identical interface for both line search and trust region Newton methods.

Downloading the Software:: The software described here is available on the Internet at `ftp://info.mcs.anl.gov/pub/petsc/petsc.tar.Z`. It is also available through the world-wide-web (WWW) at `http://www.mcs.anl.gov/petsc/petsc.html`.

We emphasize that this is research software that is provided to help you experiment with domain decomposition methods. It does have extensive man pages and examples.

Notes and References for Section A2.2

- An overview of PETSc may be found in Gropp and Smith [Gro94b].
- The standard introduction to the C programming language is Kernighan and Ritchie [Ker88].
- A short discussion of the motivations for this approach may be found in Gropp and Smith [Gro93].

A2.3 Our Standard Model for Parallel Computing

The programming model that we generally use is a single program multiple data (SPMD) distributed memory (DM) model. Codes written with PETSc will run, unmodified, on most message passing–based parallel computers, including the IBM SP, Intel Paragon and networks of workstations running MPI.

The SPMD DM model views the parallel machine as p independent processors each with its own local memory. At the present time, summer 1995, the class of processors we consider are each roughly capable of between 2 and 200 megaflops sustained floating point performance. The memories are each from 8 to 1024 megabytes in size. The number of processors p is generally between 2 and 512. Other machines that roughly fit into this category include the MasPar 2. It has up to 16,000 32 bit processors, each with up to 256 kilobytes of local memory. The MasPar machines only have a single instruction stream, which means that each processor must perform very similar operations at the same time.

Writing a parallel application code in the SPMD DM model generally consists of writing one computer program which runs with (independent) instruction streams on each processor. Communication between processors is performed by using layered software libraries which allow the programmer both to move data explicitly between memories and to construct parallel data structures that conceal the explicit communication from the application programmer. The programming language itself need not have explicit commands or data structures for parallel programming.

The explicit communication is achieved by using message passing. In an ideal situation, message passing is taken care of by the library developer, who provides the application programmer with one (or several) easy-to-use interface to parallel data structure(s).

Notes and References for Section A2.3

- There are several implementations of MPI available, which may all be accessed via the WWW at `http://www.mcs.anl.gov/mpi`. The Argonne/Mississippi State University free implementation may also be accessed by anonymous ftp at `info.mcs.anl.gov` in the directory `pub/mpi`.

References

[Arg59] J. H. Argyris and S. Kelsey. The analysis of fuselages of arbitrary cross-section and taper, Part I. *Aircraft Eng.*, 31, 1959.

[Arg60] J. H. Argyris and S. Kelsey. The analysis of fuselages of arbitrary cross-section and taper, Part II. *Aircraft Eng.*, 33, 1960.

[Axe94] Owe Axelsson. *Iterative Solution Methods*. Cambridge University Press, Cambridge, 1994.

[Bab91] Ivo Babuška, Alan Craig, Jan Mandel, and Juhani Pitkäranta. Efficient preconditioning for the p-version finite element method in two dimensions. *SIAM J. Numer. Anal.*, 28(3):624–661, 1991.

[Bak66] N. S. Bakhvalov. On the convergence of a relaxation method with natural constraints on the elliptic operator. *USSR Comput. Math. and Math. Phys.*, 6:101–135, 1966.

[Ban88] Randolph E. Bank, Todd F. Dupont, and Harry Yserentant. The hierarchical basis multigrid method. *Numer. Math.*, 52:427–458, 1988.

[Ban90] Randolph E. Bank. *PLTMG: A Software Package for Solving Elliptic Partial Differential Equations. Users' Guide 7.0*. SIAM, Philadelphia, 1990.

[Ban95] Randolph E. Bank and Jinchao Xu. The hierarchical basis multigrid method and incomplete LU decomposition. In David F. Keyes and Jinchao Xu, editors, *Seventh International Conference of Domain Decomposition Methods in Scientific and Engineering Computing*, pages 163–174. AMS, Providence, RI, 1995.

[Bel73] K. Bell, B. Hatlestad, O. E. Hansteen, and Per O. Araldsen. *NORSAM, a programming system for the finite element method. Users manual, Part 1, General description*. NTH, Trondheim, 1973.

[Ber68] P. G. Bergan and E. Ålstedt. A programming system for finite element problems. Technical report, The Norwegian Institute of Technology, Division of Structural Mechanics, 1968. In Norwegian.

[Bjø80] Petter E. Bjørstad. *Numerical solution of the biharmonic equation*. Ph.D. thesis, Stanford University, Stanford, CA, 1980.

[Bjø84] Petter E. Bjørstad and Olof B. Widlund. Solving elliptic problems on regions partitioned into substructures. In Garrett Birkhoff and Arthur Schoenstadt, editors, *Elliptic Problem Solvers II*, pages 245–256. Academic Press, New York, 1984.

[Bjø86] Petter E. Bjørstad and Olof B. Widlund. Iterative methods for the solution of elliptic problems on regions partitioned into substructures. *SIAM J. Numer. Anal.*, 23(6):1093–1120, 1986.

[Bjø87] Petter E. Bjørstad. A large scale, sparse, secondary storage, direct linear equation solver for structural analysis and its implementation on vector and parallel architectures. *J. Parallel Comput.*, 5, 1987.

[Bjø88] Petter E. Bjørstad and Anders Hvidsten. Iterative methods for substructured elasticity problems in structural analysis. In Roland Glowinski, Gene H. Golub, Gérard A. Meurant, and Jacques Périaux, editors, *Domain Decomposition Methods for Partial Differential Equations*, pages 301–312. SIAM, Philadelphia, 1988.

[Bjø89] Petter E. Bjørstad. Multiplicative and additive Schwarz methods: Convergence in the 2 domain case. In Tony F. Chan, Roland Glowinski, Jacques Périaux, and Olof B. Widlund, editors, *Domain Decomposition Methods*, pages 147–159. SIAM, Philadelphia, 1989.

[Bjø89a] Petter E. Bjørstad and Olof B. Widlund. To overlap or not to overlap: A note on a domain decomposition method for elliptic problems. *SIAM J. Sci. Stat. Comput.*, 10(5):1053–1061, 1989.

[Bjø91] Petter E. Bjørstad and Jan Mandel. On the spectra of sums of orthogonal projections with applications to parallel computing. *BIT*, 31:76–88, 1991.

[Bjø92] Petter E. Bjørstad and Morten Skogen. Domain decomposition algorithms of Schwarz type, designed for massively parallel computers. In David E. Keyes, Tony F. Chan, Gérard A. Meurant, Jeffrey S. Scroggs, and Robert G. Voigt, editors, *Fifth International Symposium on Domain Decomposition Methods for Partial Differential Equations*, pages 362–375. SIAM, Philadelphia, 1992.

[Bor93] Folkmar Bornemann and Harry Yserentant. A basic norm equivalence for the theory of multilevel methods. *Numer. Math.*, 64(4):455–476, 1993.

[Bou89] Jean-François Bourgat, Roland Glowinski, Patrick Le Tallec, and Marina Vidrascu. Variational formulation and algorithm for trace operator in domain decomposition calculations. In Tony F. Chan, Roland Glowinski, Jacques Périaux, and Olof B. Widlund, editors, *Domain Decomposition Methods*, pages 3–16. SIAM, Philadelphia, 1989.

[Bra66] James H. Bramble. A second order finite difference analogue of the first biharmonic boundary value problem. *Numer. Math.*, 9:236–249, 1966.

[Bra72] A. Brandt. Multi-level adaptive techniques (MLAT) for fast numerical solution to boundary value problems. In *Proceeding of the 3rd International Conference on Numerical Method in Fluid Mechanics*, pages 82–89. Springer, Berlin, 1972.

[Bra77] A. Brandt. Multi-level adaptive solutions to boundary value problems. *Numer. Math.*, 31:333–390, 1977.

[Bra86] James H. Bramble, Joseph E. Pasciak, and Alfred H. Schatz. The construction of preconditioners for elliptic problems by substructuring, I. *Math. Comp.*, 47(175):103–134, 1986.

[Bra87] James H. Bramble, Joseph E. Pasciak, and Alfred H. Schatz. The construction of preconditioners for elliptic problems by substructuring, II. *Math. Comp.*, 49:1–16, 1987.

[Bra88] James H. Bramble, Joseph E. Pasciak, and Alfred H. Schatz. The construction of preconditioners for elliptic problems by substructuring, III. *Math. Comp.*, 51:415–430, 1988.

[Bra89] James H. Bramble, Joseph E. Pasciak, and Alfred H. Schatz. The construction of preconditioners for elliptic problems by substructuring, IV. *Math. Comp.*, 53:1–24, 1989.

[Bra90] James H. Bramble, Joseph E. Pasciak, and Jinchao Xu. Parallel multilevel preconditioners. *Math. Comp.*, 55:1–22, 1990.

[Bra91a] James H. Bramble, Joseph E. Pasciak, Junping Wang, and Jinchao Xu. Convergence estimates for multigrid algorithms without regularity assumptions. *Math. Comp.*, 57(195):23–45, 1991.

[Bra91b] James H. Bramble, Joseph E. Pasciak, Junping Wang, and Jinchao Xu. Convergence estimates for product iterative methods with applications to domain decomposition. *Math. Comp.*, 57(195):1–21, 1991.

[Bra91c] James H. Bramble and Jinchao Xu. Some estimates for a weighted L^2 projection. *Math. Comp.*, 56:463–476, 1991.

[Bra93] James H. Bramble. *Multigrid Methods*. Pitman Research Notes in Mathematics Series No. 294. Longman Scientific & Technical, 1993.

[Bra93a] James H. Bramble, Zbigniew Leyk, and Joseph E. Pasciak. Iterative schemes for nonsymmetric and indefinite elliptic boundary value problems. *Math. Comp.*, 60:1–22, 1993.

[Bre79] Daniel Brelaz. New methods to color the vertices of a graph. *Communications of the ACM*, 22:251–256, 1979.

[Bri87] William Briggs. *A Multigrid Tutorial*. SIAM, Philadelphia, 1987.

[Bun85] J. R. Bunch. Stability of methods for solving Toeplitz systems of equations. *SIAM J. Sci. Stat. Comp.*, 6:349–364, 1985.

[Cai90] Xiao-Chuan Cai. An additive Schwarz algorithm for nonselfadjoint elliptic equations. In Tony F. Chan, Roland Glowinski, Jacques Périaux, and Olof B. Widlund, editors, *Third International Symposium on Domain Decomposition Methods for Partial Differential Equations*, pages 232–244. SIAM, Philadelphia, 1990.

[Cai91] Xiao-Chuan Cai. Additive Schwarz algorithms for parabolic convection-diffusion equations. *Numer. Math.*, 60(1):41–61, 1991.

[Cai92a] Xiao-Chuan Cai, William D. Gropp, and David E. Keyes. A comparison of some domain decomposition algorithms for nonsymmetric elliptic problems. In David E. Keyes, Tony F. Chan, Gérard A. Meurant, Jeffrey S. Scroggs, and Robert G. Voigt, editors, *Fifth International Symposium on Domain Decomposition Methods for Partial Differential Equations*, pages 224–235. SIAM, Philadelphia, 1992.

[Cai92b] Xiao-Chuan Cai, William D. Gropp, and David E. Keyes. Convergence rate estimate for a domain decomposition method. *Numer. Math.*, 61:153–169, 1992.

[Cai92c] Xiao-Chuan Cai and Olof B. Widlund. Domain decomposition algorithms for indefinite elliptic problems. *SIAM J. Sci. Statist. Comput.*, 13(1):243–258, January 1992.

[Cai93] Xiao-Chuan Cai. An optimal two-level overlapping domain decomposition method for elliptic problems in two and three dimensions. *SIAM J. Sci. Comp.*, 14:239–247, January 1993.

[Cai93a] Xiao-Chuan Cai and Olof B. Widlund. Multiplicative Schwarz algorithms for some nonsymmetric and indefinite problems. *SIAM J. Numer. Anal.*, 30(4):936–952, August 1993.

[Cai94] Xiao-Chuan Cai. Multiplicative Schwarz methods for parabolic problems. *SIAM J. Sci. Comput.*, 15(3):587–603, 1994.

[Cai94a] Xiao-Chuan Cai, William D. Gropp, and David E. Keyes. A comparison of some domain decomposition and ILU preconditioned iterative methods for nonsymmetric elliptic problems. *Numer. Lin. Alg. Appl.*, 1(5):477–504, 1994.

[Cai95] Xiao-Chuan Cai. The use of pointwise interpolation in domain decomposition methods. *SIAM J. Sci. Comput.*, 16(1), 1995.

[Cha87] Tony F. Chan. Analysis of preconditioners for domain decomposition. *SIAM J. Numer. Anal.*, 24(2):382–390, 1987.

[Cha87a] Tony F. Chan and Diana C. Resasco. Analysis of domain decomposition preconditioners on irregular regions. In R. Vichnevetsky and R. Stepleman, editors, *Advances in Computer Methods for Partial Differential Equations*, pages 317–322. IMACS, 1987.

[Cha89] Tony F. Chan, Roland Glowinski, Jacques Périaux, and Olof B. Widlund, editors. *Domain Decomposition Methods*, SIAM, Philadelphia, 1989.

[Cha90a] Tony F. Chan, Roland Glowinski, Jacques Périaux, and Olof B. Widlund, editors. *Third International Symposium on Domain Decomposition Methods for Partial Differential Equations*. SIAM, Philadelphia, 1990.

[Cha90b] Tony F. Chan and David E. Keyes. Interface preconditioning for domain-decomposed convection-diffusion operators. In Tony F. Chan, Roland Glowinski, Jacques Périaux, and Olof B. Widlund, editors, *Third International Symposium on Domain Decomposition Methods for Partial Differential Equations*, pages 245–262. SIAM, Philadelphia, 1990.

[Cha91a] Tony F. Chan and Thomas Y. Hou. Eigendecomposition of domain decomposition interface operators for constant coefficient elliptic problems. *SIAM J. Sci. Stat. Comput.*, 12:1471–1479, 1991.

[Cha91b] Tony F. Chan and Tarek P. Mathew. An application of the probing technique to the vertex space method in domain decomposition. In Roland Glowinski, Yuri A. Kuznetsov, Gérard A. Meurant, Jacques Périaux, and Olof B. Widlund, editors, *Fourth International Symposium on Domain Decomposition Methods for Partial Differential Equations*, pages 101–111. SIAM, Philadelphia, 1991.

[Cha92a] Tony F. Chan and Danny Goovaerts. On the relationship between overlapping and nonoverlapping domain decomposition methods. *SIAM J. Matrix Anal. Appl.*, 13:663–670, 1992.

[Cha92b] Tony F. Chan and Tarek P. Mathew. The interface probing technique in domain decomposition. *SIAM J. Matrix Anal. and Appl.*, 13(1):212–238, 1992.

[Cha94a] Tony F. Chan and Tarek P. Mathew. Domain decomposition algorithms. In *A. Acta Numerica 1994*, pages 61–143. Cambridge University Press, Cambridge, 1994.

[Cha94b] Tony F. Chan, Tarek P. Mathew, and Jian-Ping Shao. Efficient variants of the vertex space domain decomposition algorithm. *SIAM J. Sci. Comput.*, 15(6), 1994.

[Cha94c] Tony F. Chan and Jian-Ping Shao. Optimal coarse grid size in domain decomposition. *J. Comp. Math.*, 12(4):291–297, 1994.

[Cha94d] Tony F. Chan, Barry F. Smith, and Jun Zou. Overlapping Schwarz methods on unstructured meshes using non-matching coarse grids. Technical Report 94-8, UCLA, Dept. of Mathematics, 1994. To appear in *Numer. Math.*

[Cha95a] Tony F. Chan and Jian-Ping Shao. Parallel complexity of domain decomposition methods and optimal coarse grid sizes. *Parallel Comput.*, 21:7–16, 1995.

[Cha95b] Tony F. Chan and Barry F. Smith. Multigrid and domain decomposition on unstructured grids. In David F. Keyes and Jinchao Xu, editors, *Seventh International Conference of Domain Decomposition Methods in Scientific and Engineering Computing*. AMS, Providence, RI, 1995. A revised version of this paper has appeared in *ETNA*, 2:171–182, December 1994.

[Cia78] Philippe G. Ciarlet. *The Finite Element Method for Elliptic Problems*. North-Holland, Amsterdam, 1978.

[Cia95] Patrick Ciarlet Jr., Francoise Lamour, and Barry F. Smith. On the influence of the partitioning schemes on the efficiency of overlapping domain decomposition methods. In *Proceedings of the Fifth Symposium on the Frontiers of Massively Parallel Computation*, pages 375–383. IEEE Press, Los Alamitos, CA, 1995.

[Con76] Paul Concus, Gene H. Golub, and Diana O'Leary. A generalized conjugate gradient method for the numerical solution of elliptic partial differential equations. In J. R. Bunch and D. J. Rose, editors, *Sparse Matrix Computations*. Academic Press, New York, 1976.

[Cot74] R. W. Cottle. Manifestations of the Schur complement. *Lin. Alg. and Its Applic.*, 8:189–211, 1974.

[Cow92] Lawrence Cowsar, Jan Mandel, and Mary F. Wheeler. Balancing domain decomposition for mixed problems in oil reservoir simulation. Presented at the 6th International Symposium on Domain Decomposition Methods, Como, Italy, 1992.

[Cow93] Lawrence C. Cowsar, Jan Mandel, and Mary F. Wheeler. Balancing domain decomposition for mixed finite elements. Technical Report TR93-08, Department of Mathematical Sciences, Rice University, March 1993.

[Cra76] H. L. Crane Jr., Norman E. Gibbs, William G. Poole Jr., and Paul K. Stockmeyer. Algorithm 508, matrix bandwith and profile reduction. *ACM Trans. Math. Software*, 2(4), December 1976.

[Cur74] A. R. Curtis, M. J. D. Powell, and J. K. Reid. On the estimation of sparse Jacobian matrices. *J. Inst. Math. Appl.*, 13:117–120, 1974.

[De R91] Yann-Hervé De Roeck and Patrick Le Tallec. Analysis and test of a local domain decomposition preconditioner. In Roland Glowinski, Yuri Kuznetsov, Gérard Meurant, Jacques Périaux, and Olof B. Widlund, editors, *Fourth International Symposium on Domain Decomposition Methods for Partial Differential Equations*, pages 112–128. SIAM, Philadelphia, 1991.

[Deu89] P. Deuflhard, P. Leinen, and Harry Yserentant. Concepts of an adaptive hierarchical finite element code. *IMPACT*, 1:3–35, 1989.

[Deu95] P. Deuflhard. Cascadic conjugate gradient methods for elliptic partial equations I. algorithm and numerical results. In David F. Keyes and Jinchao Xu, editors, *Seventh International Conference of Domain Decomposition Methods in Scientific and Engineering Computing*, pages 29–42. AMS, Providence, RI, 1995.

[Dih84] Q. V. Dihn, Roland Glowinski, and Jacques Périaux. Solving elliptic problems by domain decomposition methods with applications. In Garrett Birkhoff and Arthur Schoenstadt, editors, *Elliptic Problem Solvers II*, pages 395–426. Academic Press, New York, 1984.

[Dry81] Maksymilian Dryja. An algorithm with a capacitance matrix for a variational-difference scheme. In G. I. Marchuk, editor, *Variational-Difference Methods in Mathematical Physics*, pages 63–73. USSR Academy of Sciences, Novosibirsk, 1981.

[Dry87] Maksymilian Dryja and Olof B. Widlund. An additive variant of the Schwarz alternating method for the case of many subregions. Technical Report 339, also Ultracomputer Note 131, Department of Computer Science, Courant Institute, 1987.

[Dry88] Maksymilian Dryja. A method of domain decomposition for 3-D finite element problems. In Roland Glowinski, Gene H. Golub, Gérard A. Meurant, and Jacques Périaux, editors, *First International Symposium on Domain Decomposition Methods for Partial Differential Equations*, pages 43–61. SIAM, Philadelphia, 1988.

[Dry89] Maksymilian Dryja and Olof B. Widlund. Some domain decomposition algorithms for elliptic problems. In Linda Hayes and David Kincaid, editors, *Iterative Methods for Large Linear Systems*, pages 273–291. Academic Press, San Diego, CA, 1989.

[Dry90] Maksymilian Dryja and Olof B. Widlund. Towards a unified theory of domain decomposition algorithms for elliptic problems. In Tony F. Chan, Roland Glowinski, Jacques Périaux, and Olof B. Widlund, editors, *Third International Symposium on Domain Decomposition Methods for Partial Differential Equations*, pages 3–21. SIAM, Philadelphia, 1990.

[Dry91] Maksymilian Dryja and Olof B. Widlund. Multilevel additive methods for elliptic finite element problems. In Wolfgang Hackbusch, editor, *Parallel Algorithms for Partial Differential Equations, Proceedings of the Sixth GAMM-Seminar*, Kiel, January 19–21, 1990. Vieweg, Braunschweig, Germany, 1991.

[Dry92] Maksymilian Dryja and Olof B. Widlund. Additive Schwarz methods for elliptic finite element problems in three dimensions. In David E. Keyes, Tony F. Chan, Gérard A. Meurant, Jeffrey S. Scroggs, and Robert G. Voigt, editors, *Fifth International Symposium on Domain Decomposition Methods for Partial Differential Equations*, pages 3–18. SIAM, Philadelphia, 1992.

[Dry94a] Maksymilian Dryja, Barry F. Smith, and Olof B. Widlund. Schwarz analysis of iterative substructuring algorithms for elliptic problems in three dimensions. *SIAM J. Numer. Anal.*, 31(6):1662–1694, December 1994.

[Dry94b] Maksymilian Dryja and Olof B. Widlund. Domain decomposition algorithms with small overlap. *SIAM J. Sci. Comput.*, 15(3):604–620, May 1994.

[Dry95] Maksymilian Dryja and Olof B. Widlund. Schwarz methods of Neumann-Neumann type for three-dimensional elliptic finite element problems. *Comm. Pure Appl. Math.* 48:121–155, 1995.

[Duf86] I. S. Duff, A. M. Erisman, and J. K. Reid. *Direct Methods for Sparse Matrices*. Oxford University Press, Oxford, 1986.

[Eij91] Victor Eijkhout and Panayot Vassilevski. The role of the strengthened Cauchy-Buniakowskii-Schwarz inequality in multilevel methods. *SIAM Review*, 33(3):405–419, 1991.

[Elm95] Howard C. Elman and Xuejun Zhang. Algebraic analysis of the hierarchical basis preconditioner. *SIAM J. Matrix Anal. Appl.*, 16(1):192–205, January 1995.

[Far94] Charbel Farhat and François-Xavier Roux. Implicit parallel processing in structural mechanics. In J. Tinsley Oden, editor, *Computational Mechanics Advances*, volume 2 (1), pages 1–124. North-Holland, Amsterdam, 1994.

[Fed61] R. P. Fedorenko. The speed of convergence of one iterative process. *USSR Comput. Math. and Math. Phys.*, 1:1092–1096, 1961.

[Fis94] Paul. F. Fischer. Domain decomposition methods for large scale parallel Navier-Stokes calculations. In Alfio Quarteroni, Jacques Périaux, and Yuri A. Kuznetsov, Olof Widlund, editors, *Domain Decomposition Methods in Science and Engineering*. AMS, Providence, RI, 1994.

[Fos92] Ian Foster, William D. Gropp, and Rick Stevens. The parallel scalability of the spectral transform method. *Monthly Weather Review*, 120:835–850, 1992.

[Fos95] Ian Foster. *Designing and Building Parallel Programs*. Addison-Wesley, Reading, MA, 1995.

[Fre92] Roland Freund, Gene H. Golub, and Noel Nachtigal. Iterative solution of linear systems, In A. Iserles, editor, *Acta Numerica 1992*, pages 57–100. Cambridge University Press, Cambridge, 1992.

[Geo81] Alan George and Joseph Liu. *Computer Solution of Large Sparse Positive Definite Systems*. Prentice-Hall, Englewood Cliffs, NJ, 1981.

[Gib76] Norman E. Gibbs. Algorithm 509, a hybrid profile reduction algorithm. *ACM Transactions on Mathematical Software*, 2(4), December 1976.

[Gil87] John R. Gilbert and Earl Zmijewski. A parallel graph partitioning algorithm for a message-passing multiprocessor. *Int. J. Parallel Programming*, 16:427–448, 1987.

[Glo88a] Roland Glowinski, Gene H. Golub, Gérard A. Meurant, and Jacques Périaux, editors. *Domain Decomposition Methods for Partial Differential Equations*, *Proceedings of the First International Symposium on Domain Decomposition Methods for Partial Differential Equations*, Paris, January 1987. SIAM, Philadelphia, 1988.

[Glo88b] Roland Glowinski and Mary F. Wheeler. Domain decomposition and mixed finite elements for elliptic problems. In Roland Glowinski, Gene H. Golub, Gérard A. Meurant, and Jacques Périaux, editors, *First International Symposium on Domain Decomposition Methods for Partial Differential Equations*, pages 144–172. SIAM, Philadelphia, 1988.

[Glo91] Roland Glowinski, Yuri A. Kuznetsov, Gérard A. Meurant, Jacques Périaux, and Olof B. Widlund, editors. *Fourth International Symposium on Domain Decomposition Methods for Partial Differential Equations*, Moscow, May 21–25, 1990. SIAM, Philadelphia, 1991.

[Gold92] Mark Goldberg and Thomas Spencer. An efficient parallel algorithm that finds independent sets of guaranteed size. Technical report, Rensselaer Polytechnic Institute, 1992.

[Golu84] Gene H. Golub and D. Mayers. The use of preconditioning over irregular regions. In R. Glowinski and J. L. Lions, editors, *Computing Methods in Applied Sciences and Engineering, VI*, pages 3–14. Proceedings of a Conference held in Versailles, France, December 12–16, 1983. North Holland, Amsterdam, 1984.

[Golu88] Gene H. Golub and Michael L. Overton. The convergence of inexact Chebyshev and Richardson iterative methods for solving linear systems. *Numer. Math.*, 53:571–593, 1988.

[Gre89] Anne Greenbaum and G. H. Rodrigue. Optimal preconditioners of a given sparsity pattern. *BIT*, 29:610–634, 1989.

[Gri94a] Michael Griebel. Multilevel algorithms considered as iterative methods on semidefinite systems. *SIAM J. Sci. Comput.*, 15(3):547–565, May 1994.

[Gri94b] Michael Griebel. *Multilevelmethoden als Iterationsverfahren über Erzeugendensystem*. B. G. Teubner, Stuttgart, 1994.

[Gri95] Michael Griebel and Peter Oswald. On the abstract theory of additive and multiplicative Schwarz algorithms. *Numer. Math.*, 70:163–180, 1995.

[Gro88] William D. Gropp and David E. Keyes. Complexity of parallel implementation of domain decomposition techniques for elliptic partial differential equations. *SIAM J. Stat. Sci. Comput.*, 9(2):312–326, 1988.

[Gro90] William D. Gropp and E. Smith. Computational fluid dynamics on parallel processors. *Computers and Fluids*, 18:289–304, 1990.

[Gro91] William D. Gropp and David E. Keyes. Parallel domain decomposition and the solution of nonlinear systems of equations. In Roland Glowinski, Yuri A. Kuznetsov, Gérard A. Meurant, Jacques Périaux, and Olof B. Widlund, editors, *Fourth International Symposium on Domain Decomposition Methods for Partial Differential Equations*, pages 373–381. SIAM, Philadelphia, 1991.

[Gro92] William D. Gropp. Parallel computing and domain decomposition. In David E. Keyes, Tony F. Chan, Gérard A. Meurant, Jeffrey S. Scroggs, and Robert G. Voigt, editors, *Fifth International Symposium on Domain Decomposition Methods for Partial Differential Equations*, pages 349–361. SIAM, Philadelphia, 1992.

[Gro92a] William D. Gropp and David E. Keyes. Domain decomposition methods in computational fluid dynamics. *Int. J. Numer. Meth. Fluids*, 14:147–165, 1992.

[Gro93] William D. Gropp and Barry F. Smith. The design of data-structure-neutral libraries for the iterative solution of sparse linear systems. Technical Report MCS-P356-0393, Argonne National Laboratory, 1993.

[Gro94a] William D. Gropp, Ewing Lusk, and Anthony Skjellum. *Using MPI: Portable Parallel Programming with the Message Passing Interface*. MIT Press, Cambridge, MA, 1994.

[Gro94b] William D. Gropp and Barry F. Smith. Scalable, extensible, and portable numerical libraries. In *Proceedings of Scalable Parallel Libraries Conference*, pages 87–93. IEEE, Los Alamitos, CA, 1994.

[Haa90] Gundolf Haase, Ulrich Langer, and Arnd Meyer. A new approach to the Dirichlet domain decomposition method. In S. Hengst, editor, *Fifth Multigrid Seminar, Eberswalde 1990*, pages 1–59. Report R-MATH-09/90, Karl–Weierstrass–Institut, Berlin, 1990.

[Hac85] Wolfgang Hackbusch. *Multigrid Methods and Applications*. Springer, Berlin, 1985.

[Hac94] Wolfgang Hackbusch. *Iterative Solution of Large Sparse Linear Systems of Equations*. Springer, Berlin, 1994.

[Hen93a] Bruce Hendrickson and Robert Leland. The Chaco user's guide, version 1.0. Technical Report SAND 93-2339, Sandia National Laboratories, October 1993.

[Hen93b] Bruce Hendrickson and Robert Leland. An improved spectral load balancing method. In *Proceedings of the Sixth SIAM Conference on Parallel Processing for Scientific Computation*, pages 953–961, March 1993.

[Hen95] Bruce Hendrickson and Robert Leland. An improved spectral graph partitioning algorithm for mapping parallel computations. *SIAM J. Sci. Comput.*, 16(2), 1995.

[Hvi90] Anders Hvidsten. On the parallelization of a finite element structural analysis program. Ph.D. thesis, University of Bergen, Computer Science Department, Norway, 1990.

[Hvi93] Anders Hvidsten and Jon Brækhus. Benchmarking of the parallel SESTRA program on a cluster of IBM RS6000 computers. Technical Report, Det Norske Veritas SESAM as, P.O. Box 300, N-1322 Høvik, Norway, 1993.

[Joh87] Claes Johnson. *Numerical Solutions of Partial Differential Equations by the Finite Element Method*. Cambridge University Press, Cambridge, 1987.

[Jon93] Mark T. Jones and Paul E. Plassmann. A parallel graph coloring heuristic. *SIAM J. Sci. Comput.*, 14(3):654–669, 1993.

[Jon94] Mark T. Jones and Paul E. Plassmann. Parallel algorithms for the adaptive refinement and partitioning of unstructured meshes. In *Proceedings of the 1994 SHPCC*, pages 726–733. IEEE Press, Los Alamitos, CA, 1994.

[Ker70] B. W. Kernighan and S. Lin. An efficient heuristic procedure for partitioning graphs. *Bell System Tech. J.*, pages 291–307, February 1970.

[Ker88] B. Kernighan and D. Ritchie. *The C Programming Language*. Prentice-Hall, Englewood Cliffs, NJ, 1988.

[Key87] David E. Keyes and William D. Gropp. A comparison of domain decomposition techniques for elliptic partial differential equations and their parallel implementation. *SIAM J. Sci. Stat. Comput.*, 8(2):s166–s202, 1987.

[Key90a] David E. Keyes and William D. Gropp. Decomposition techniques for the parallel solution of nonsymmetric systems of elliptic BVPs. *Appl. Numer. Math.*, 6:281–301, 1990.

[Key90b] David E. Keyes and William D. Gropp. Domain decomposition techniques for the parallel solution of nonsymmetric systems of elliptic BVPs. *Appl. Numer. Math.*, 6:281–301, 1990.

[Key92a] David E. Keyes, Tony F. Chan, Gérard A. Meurant, Jeffrey S. Scroggs, and Robert G. Voigt, editors. *Fifth International Symposium on Domain Decomposition Methods for Partial Differential Equations*, Norfolk, VA, May 6–8, 1991, SIAM, Philadelphia, 1992.

[Key92b] David E. Keyes and William D. Gropp. Domain decomposition with local mesh refinement. *SIAM J. Sci. Stat. Comput.*, 13:967–993, 1992.

[Key95] David E. Keyes and Jinchao Xu, editors. *Domain Decomposition Methods in Science and Engineering*, Proceedings of the Seventh International Conference on Domain Decomposition, October 27–30, 1993, The Pennsylvania State University. AMS, Providence, RI, 1995.

[Kho92] B. N. Khoromskij and W. L. Wendland. Spectrally equivalent preconditioners for boundary equations in substructuring techniques. *East-West J. Numer. Anal.*, 1:1–27, 1992.

[Koe94] Charles H. Koelbel, David B. Loveman, Robert S. Schreiber, Guy L. Steele Jr., and Mary E. Zosel. *The High Performance Fortran Handbook*. MIT Press, Cambridge, MA, 1994.

[Krå74] B. Kråkeland, P. I. Johansson, E. Pahle, B. Blaker, and H. F. Klem. General superelement program SESAM-69, User's Manual NV336. Technical Report, Det norske Veritas, P.O. Box 300, N-1322 Høvik, Norway, 1974.

[Kro71] L. Kronsjø and G. Dahlquist. On the design of nested iterations for elliptic difference equations. *BIT*, 11:63–71, 1971.

[Kuz93] Yuri Kuznetsov, Petri Manninen, and Yuri Vassilevski. On numerical experiments with Neumann-Neumann and Neumann-Dirichlet domain decomposition preconditioners. Technical Report, University of Jyväskylä, 1993.

[Le T91] Patrick Le Tallec, Yann-Hervé De Roeck, and Marina Vidrascu. Domain-decomposition methods for large linearly elliptic three dimensional problems. *J. Comput. Appl. Math.*, 34, 1991.

[Le T94] Patrick Le Tallec. Domain decomposition methods in computational mechanics. In J. Tinsley Oden, editor, *Computational Mechanics Advances*, volume 1 (2), pages 121–220. North-Holland, New York, 1994.

[Lew89] John G. Lewis, Barry W. Peyton, and Alex Pothen. A fast algorithm for reordering sparse matrices for parallel factorization. *SIAM J. Sci. Stat. Comput.*, 10(6):1146–1173, 1989.

[Lio78] Pierre Louis Lions. Interprétation stochastique de la méthode alternée de Schwarz. *C. R. Acad. Sci. Paris*, 268:325–328, 1978.

[Lio88] Pierre Louis Lions. On the Schwarz alternating method. I. In Roland Glowinski, Gene H. Golub, Gérard A. Meurant, and Jacques Périaux, editors, *First International Symposium on Domain Decomposition Methods for Partial Differential Equations*, pages 1–42. SIAM, Philadelphia, 1988.

[Lio89] Pierre Louis Lions. On the Schwarz alternating method. II. In Tony F. Chan, Roland Glowinski, Jacques Périaux, and Olof B. Widlund, editors, *Domain Decomposition Methods*, pages 47–70. SIAM, Philadelphia, 1989.

[Liu89] J. W. H. Liu. The minimum degree ordering with constraints. *SIAM J. Sci. Stat. Comput.*, 10(6):1136–1145, 1989.

[Liu90] J. W. H. Liu. The role of elimination trees in sparse factorization. *SIAM J. Matrix Anal. Appl.*, 11:134–172, 1990.

[Liu92] J. W. H. Liu. The multifrontal method for sparse matrix solution: Theory and practice. *SIAM Review*, 34:82–109, 1992.

[Lub86] Michael Luby. A simple parallel algorithm for the maximal independent set problem. *SIAM J. Comput.*, 15(4):1036–1053, November 1986.

[Man90a] Jan Mandel. Iterative solvers by substructuring for the p-version finite element method. *Comp. Meth. Appl. Mech. Eng.*, 80:117–128, 1990.

[Man90b] Jan Mandel. On block diagonal and Schur complement preconditioning. *Numer. Math.*, 58:79–93, 1990.

[Man90c] Jan Mandel. Two-level domain decomposition preconditioning for the p-version finite element version in three dimensions. *Int. J. Numer. Meth. Eng.*, 29:1095–1108, 1990.

[Man91] Jan Mandel and G. Scott Lett. Domain decomposition preconditioning for p-version finite elements with high aspect ratios. *Applied Numerical Analysis*, 8:411–425, 1991.

[Man92] Jan Mandel and Marian Brezina. Balancing domain decomposition for problems with large jumps in coefficients: Theory and computations in two and three dimensions. *Math. Comp.*, forthcoming.

[Man93] Jan Mandel. Balancing domain decomposition. *Comm. Numer. Meth. Eng.*, 9:233–241, 1993.

[Man94] Jan Mandel. Hybrid domain decomposition with unstructured subdomains. In Alfio Quarteroni, Yuri A. Kuznetsov, Jacques Périaux, and Olof B. Widlund, editors, *Domain Decomposition Methods in Science and Engineering: The Sixth International Conference on Domain Decomposition*, volume 157, pages 103–112. AMS, Providence, RI, 1994.

[Math89] Tarek P. Mathew. *Domain Decomposition and Iterative Refinement Methods for Mixed Finite Element Discretizations of Elliptic Problems*. Ph.D. thesis, Courant Institute of Mathematical Sciences, September 1989. Tech. Rep. 463, Department of Computer Science, Courant Institute.

[Mats85] A. M. Matsokin and Sergey V. Nepomnyaschikh. A Schwarz alternating method in a subspace. *Soviet Mathematics*, 29(10):78–84, 1985.

[Mats88] A. M. Matsokin and Sergey V. Nepomnyaschikh. Norms on the space of traces of mesh functions. *Sov. J. Numer. Anal. Math. Modeling*, 3:199–216, 1988.

[Mats89] A. M. Matsokin and Sergey V. Nepomnyaschikh. On using the bordering method for solving systems of mesh equations. *Sov. J. Numer. Anal. Math. Modeling*, 4:487–492, 1989.

[McC86] Stephen F. McCormick and J. Ruge. Unigrid for multigrid simulation. *Math. Comp.*, 41:43–62, 1986.

[McC87] Stephen F. McCormick, editor. *Multigrid Methods*. SIAM, Philadelphia, 1987.

[McC89] Stephen F. McCormick. *Multilevel Adaptive Methods for Partial Differential Equations*. SIAM, Philadelphia, 1989.

[Mil65] Keith Miller. Numerical analogs to the Schwarz alternating procedure. *Numer. Math.*, 7:91–103, 1965.

[Mor56] D. Morgenstern. Begründung des alternierenden Verfahrens durch Orthogonalprojektion. *ZAMM*, 36:7–8, 1956.

[Mun87] Hans Munthe-Kaas. The convergence rate of inexact preconditioned steepest descent algorithm for solving linear systems. Technical Report NA-87-04, Computer Science Department, Stanford University, 1987.

[Nep84] Sergey V. Nepomnyaschikh. On the application of the method of bordering for elliptic mixed boundary value problems and on the difference norms of $W_2^{1/2}(S)$. Technical Report 106, Computing Center of the Siberian Branch of the USSR Academy of Sciences, Novosibirsk, 1984. In Russian.

[Nep86] Sergey V. Nepomnyaschikh. *Domain Decomposition and Schwarz Methods in a Subspace for the Approximate Solution of Elliptic Boundary Value Problems*. Ph.D. thesis, Computing Center of the Siberian Branch of the USSR Academy of Sciences, Novosibirsk, USSR, 1986.

[Nep89] Sergey V. Nepomnyaschikh. On the application of the bordering method to the mixed boundary value problem for elliptic equations and on the mesh norms in $W_2^{1/2}(S)$. *Sov. J. Numer. Anal. Math. Modeling*, 4:493–506, 1989.

[Nep91a] Sergey V. Nepomnyaschikh. Application of domain decomposition to elliptic problems with discontinuous coefficients. In Roland Glowinski, Yuri A. Kuznetsov, Gérard A. Meurant, Jacques Périaux, and Olof B. Widlund, editors, *Fourth International Symposium on Domain Decomposition Methods for Partial Differential Equations*, pages 242–251. SIAM, Philadelphia, 1991.

[Nep91b] Sergey V. Nepomnyaschikh. Mesh theorems of traces, normalizations of function traces and their inversions. *Sov. J. Numer. Anal. Math. Modeling*, 6:1–25, 1991.

[Ong89] M. E. G. Ong. Hierarchical basis preconditioners for second order elliptic problems in three dimensions. Technical Report 89-3, Dept. of Applied Math. University of Washington, Seattle, 1989.

[Osw90] Peter Oswald. On function spaces related to finite element approximation theory. *Z. Anal. Anwendungen*, 9(1):43–64, 1990.

[Osw91] Peter Oswald. On a BPX-preconditioner for P1 elements. Technical Report Math./91/2, Friedrich Schiller Universität, Jena, Germany, 1991. (part 2).

[Osw92a] Peter Oswald. On a hierarchical basis multilevel method with nonconforming P1 elements. *Numer. Math.*, 62(2):189–212, 1992.

[Osw92b] Peter Oswald. On discrete norm estimates related to multilevel preconditioners in the finite element method. In K. Ivanov and B. Sendov, editors, *Proceedings of the International Conference on Constructive Theory of Functions, Varna 91*, pages 203–241, 1992.

[Osw94] Peter Oswald. *Multilevel Finite Element Approximation, Theory and Applications*. Teubner Skripten zur Numerik. B.G./Teubner, Stuttgart, 1994.

[Pav94] Luca F. Pavarino. Additive Schwarz methods for the p-version finite element method. *Numer. Math.*, 66(4):493–515, 1994.

[Pav94a] Luca F. Pavarino and Olof B. Widlund. Iterative substructuring methods for spectral elements in three dimensions. In M. Křížek, P. Neittaanmäki, and R. Stenberg, editors, *The Finite Element Method: Fifty Years of the Courant Element*, pages 345–355. Dekker, New York, 1994.

[Pav94b] Luca F. Pavarino and Olof B. Widlund. A polylogarithmic bound for an iterative substructuring method for spectral elements in three dimensions. Technical Report 661, Courant Institute of Mathematical Sciences, Department of Computer Science, March 1994. To appear in *SIAM J. Numer. Anal.*

[Pot90] Alex Pothen, Horst D. Simon, and Kang-Pu Liou. Partitioning sparse matrices with eigenvectors of graphs. *SIAM J. Matrix Anal. Appl.*, 11(3):430–452, July 1990.

[Pot92] Alex Pothen, Horst D. Simon, and L. Wang. Spectral nested dissection. Department of Computer Science CS-91-01, Pennsylvania State University, 1992.

[Pow79] M. J. D. Powell and PH. L. Toint. On the estimation of sparse hessian matrices. *SIAM J. Numer. Anal.*, 16(6):1060–1074, 1979.

[Prz63] J. S. Przemieniecki. Matrix structural analysis of substructures. *Am. Inst. Aero. Astro. J.*, 1:138–147, 1963.

[Prz85] J. S. Przemieniecki. *Theory of Matrix Structural Analysis*. Dover, New York, 1985. Reprint of McGraw-Hill, 1968.

[Qua94] Alfio Quarteroni, Yuri A. Kuznetsov, Jacques Périaux, and Olof B. Widlund, editors. *Domain Decomposition Methods in Science and Engineering: The Sixth International Conference on Domain Decomposition*, Como, Italy, June 15–19, 1992, volume 157. AMS, Providence, RI, 1994.

[Rei84] J. K. Reid. Treesolve, a Fortran package for solving large sets of linear finite-element equations. In Bjørn Enquist and Tom Smedsaas, editors, *PDE SOFTWARE: Modules, Interfaces and Systems*, pages 1–17, Elsevier Science, North Holland, Amsterdam, 1984.

[Roi89a] R. Roitzsch. Kaskade programmer's manual. Technical Report TR 89-5, Konrad-Zuse-Zentrum für Informationstechnik, 1989.

[Roi89b] R. Roitzsch. Kaskade user's manual. Technical Report TR 89-4, Konrad-Zuse-Zentrum für Informationstechnik, 1989.

[Rüd93] Ulrich Rüde. *Mathematical and Computational Techniques for Multilevel Adaptive Methods*. SIAM, Philadelphia, 1993.

[Saa93] Youcef Saad. A flexible inner-outer preconditioned GMRES algorithm. *SIAM J. Sci. Stat. Comput.*, 14:461–469, 1993.

[Saa95] Youcef Saad. *Iterative Methods for Sparse Linear Systems*. PWS Kent, 1995.

[Sav91] John E. Savage and Markus G. Wloka. Parallelism in graph-partitioning. *J. Parallel and Distributed Comput.*, 13:257–272, 1991.

[Sch74] Alfred H. Schatz. An observation concerning Ritz-Galerkin methods with indefinite bilinear forms. *Math. Comp.*, 28(128):959–962, 1974.

[Sch82] Robert Schreiber. A new implementation of sparse gaussian elimination. *ACM Trans. on Math. Software*, 8:256–276, 1982.

[Sch90] H. A. Schwarz. *Gesammelte Mathematische Abhandlungen*, volume 2, pages 133–143. Springer, Berlin, 1890. First published in *Vierteljahrsschrift Naturforsch. Ges. Zürich*, 15:272–286, 1870.

[Sko92] Morten D. Skogen. *Schwarz Methods and Parallelism*. Ph.D. thesis, Department of Informatics, University of Bergen, Norway, February 1992.

[Smi90] Barry F. Smith. *Domain Decomposition Algorithms for the Partial Differential*

Equations of Linear Elasticity. Ph.D. thesis, Courant Institute of Mathematical Sciences, September 1990. Tech. Rep. 517, Department of Computer Science, Courant Institute.

[Smi90a] Barry F. Smith and Olof B. Widlund. A domain decomposition algorithm using a hierarchical basis. *SIAM J. Sci. Stat. Comput.*, 11(6):1212–1220, 1990.

[Smi91] Barry F. Smith. A domain decomposition algorithm for elliptic problems in three dimensions. *Numer. Math.*, 60(2):219–234, 1991.

[Smi92] Barry F. Smith. An optimal domain decomposition preconditioner for the finite element solution of linear elasticity problems. *SIAM J. Sci. Stat. Comput.*, 13(1):364–378, January 1992.

[Smi93] Barry F. Smith. A parallel implementation of an iterative substructuring algorithm for problems in three dimensions. *SIAM J. Sci. Comput.*, 14(2):406–423, March 1993.

[Smi94] Barry F. Smith. Extensible PDE solvers package users manual. Technical Report ANL-94/40, Argonne National Laboratory, 1994. Available via anonymous ftp at info.mcs.anl.gov in the directory pub/petsc.

[Smi95] Barry F. Smith, William D. Gropp, and Lois Curfman McInncs. *PETSc 2.0 Users Manual*, Tech. Rep. ANL-95/11, Argonne National Laboratory, 1995. (Available via ftp://www.mcs.anl/pub/petsc/manual.ps)

[Sou35a] R. V. Southwell. Stress-calculation in frameworks by the method of "systematic relaxation of constraints," parts I and II. *Proc. Roy. Soc. Edinburgh Sect. A*, 151:57–91, 1935.

[Sou35b] R. V. Southwell. Stress-calculation in frameworks by the method of "systematic relaxation of constraints," part III. *Proc. Roy. Soc. Edinburgh Sect. A*, 153:41–76, 1935.

[Sta77] G. Starius. Composite mesh difference methods for elliptic boundary value problems. *Numer. Math.*, 28:243–258, 1977.

[Str72] Gilbert Strang. Approximation in the finite element method. *Numer. Math.*, 19:81–98, 1972.

[Str73] Gilbert Strang and George J. Fix. *An Analysis of the Finite Element Method*. Prentice-Hall, Englewood Cliffs, NJ, 1973.

[Tan92] Wei Pai Tang. Generalized Schwarz splittings. *SIAM J. Sci. Stat. Comput.*, 13:573–595, 1992.

[Ton91] Charles H. Tong, Tony F. Chan, and C. C. Jay Kuo. A domain decomposition preconditioner based on a change to a multilevel nodal basis. *SIAM J. Sci. Stat. Comput.*, 12(6):1486–1495, 1991.

[Var62] Richard S. Varga. *Matrix Iterative Analysis*. Prentice-Hall, Englewood Cliffs, NJ, 1962.

[Wes92] Peter Wesseling. *An Introduction to Multigrid Methods*. Wiley, New York, 1992.

[Xu89] Jinchao Xu. *Theory of Multilevel Methods*. Ph.D. thesis, Cornell University, May 1989.

[Xu92] Jinchao Xu. Iterative methods by space decomposition and subspace correction. *SIAM Review*, 34:581–613, December 1992.

[Yse86a] Harry Yserentant. Hierarchical bases give conjugate gradient type methods a multigrid speed of convergence. *Appl. Math. Comp.*, 19:347–358, 1986.

[Yse86b] Harry Yserentant. On the multi-level splitting of finite element spaces. *Numer. Math.*, 49:379–412, 1986.

[Yse86c] Harry Yserentant. On the multi-level splitting of finite element spaces for indefinite elliptic boundary value problems. *SIAM J. Numer. Anal.*, 23:581–595, 1986.

[Yse90] Harry Yserentant. Two preconditioners based on the multi-level splitting of finite element spaces. *Numer. Math.*, 58(2):163–184, 1990.

[Yse93] Harry Yserentant. Old and new convergence proofs for multigrid methods. In A. Iserles, editor, *Acta Numerica 1993*, pages 285–326. Cambridge University Press, Cambridge, 1993.

[Zer90] Janez Zerovnik. A parallel variant of a heuristical algorithm for graph colouring. *Parallel Comput.*, 13:95–100, 1990.

[Zha91] Xuejun Zhang. *Studies in Domain Decomposition: Multilevel Methods and the Biharmonic Dirichlet Problem*. Ph.D. thesis, Courant Institute, New York University, September 1991.

[Zha92] Xuejun Zhang. Multilevel Schwarz methods. *Numer. Math.*, 63(4):521–539, 1992.

[Zha94] Xuejun Zhang. Multilevel Schwarz methods for the biharmonic Dirichlet problem. *SIAM J. Sci. Comput.*, 15(3):621–644, 1994.

Index